CASS LIBRARY OF SCIENCE CLASSICS
No. 10

General Editor: Dr. L. L. LAUDAN, University College London

THE
CELESTIAL WORLDS
DISCOVER'D

THE
CELESTIAL WORLDS
DISCOVER'D

Christian Huygens

Routledge
Taylor & Francis Group
LONDON AND NEW YORK

First published by
FRANK CASS AND COMPANY LIMITED

Published 2006 by Routledge
2 Park Square, Milton Park, Abingdon, Oxfordshire OX14 4RN
711 Third Avenue, New York, NY 10017

First issued in paperback 2014

Routledge is an imprint of the Taylor and Francis Group, an informa business

First Latin edition (*Cosmotheoros*)	1698
First English edition	1698
Second Latin edition	1699
First French edition	1718
Second English edition	1722
Third English edition	1757
New impression of the First English edition	1968

ISBN 13: 978-0-7146-1602-5 (hbk)
ISBN 13: 978-0-415-76062-1 (pbk)

Publisher's Note to the 1968 Edition

This is an exact facsimile reproduction of the posthumous 1698 edition of Christian Huygen's *The Celestial Worlds Discover'd*, except that an index, prepared by the General Editor, has been added.

THE
Celeſtial Worlds
DISCOVER'D:
OR,
CONJECTURES
Concerning the
INHABITANTS,
PLANTS and PRODUCTIONS
OF THE
𝔚𝔬𝔯𝔩𝔡𝔰 𝔦𝔫 𝔱𝔥𝔢 𝔓𝔩𝔞𝔫𝔢𝔱𝔰.

Written in Latin by
CHRISTIANVS HVYGENS,
And inſcrib'd to his Brother
CONSTANTINE HVYGENS,
Late Secretary to his Majeſty K. *William.*

LONDON,
Printed for TIMOTHY CHILDE at the
White Hart at the Weſt-end of St.
Paul's Church-yard. MDCXCVIII.

TO THE

READER.

THIS Book was juſt finiſhed, and deſigned for the Preſs, when the Author, to the great loſs of the Learned World, was ſeized by a Diſeaſe that brought him to his Death. However he took care in his laſt Will of its Publication, deſiring his Brother, to whom it was writ, to take that Trouble upon him. But he was ſo taken up with Buſineſs and Removals, (as being Secretary in *Holland* to the King of *Great Britain*) that he could find no time for it till a year after the Death of the Author : When it ſo fell out, that the Printers being ſomewhat tardy, and this Gentleman dying, the Book was left without either Father or Guardian. Yet it now ventures into the Publick, in the ſame method that it was writ by the Author,

thor, and with the same Inscription
to his Brother, tho dead ; in confi-
dence that this last Piece of his will
meet with as kind a reception from the
World as all the other Works of that
Author have. 'Tis true there are not
every where Mathematical Demon-
strations ; but where they are wanting, you have probable and ingenious
Conjectures, which is the most that
can be reasonably expected in such
matters. What belongs to, or has
any thing to do with Astronomy, you
will see demonstrated, and the rest in-
geniously and shrewdly guess'd at,
from the affinity and relation of the
heavenly Bodies to the Earth. For
your farther Satisfaction read on, and
farewel.

THE

(v)

THE
PUBLISHER
TO THE
READER.

I Doubt not but I *shall incur the Cen-*
sures of learned Men for putting
this Book into English, because,
they'l say, it renders Philosophy cheap
and vulgar, and, which is worse, fur-
nishes a sort of injudicious People with
a smattering of Notions, which being
not able to make a proper use of, they per-
vert to the Injury of Religion and Sci-
ence. I confess the Allegation is too true :
but after Bishop Wilkins, Dr. Burnet,
Mr. Whifton, *and others, to say nothing*
of the antient Philosophers, who wrote in
their own Tongues ; I say after these great
Authors have treated on as learned and
abstruse Subjects in the same Language,
I

I hope their Example will be allowed a sufficient excuse for printing this Book in English.

Concerning this Edition I can say, that I have taken care to have the Cuts exactly done, and have plac'd each Figure at the Page of the Book that refers to it, which I take to be more convenient to the Reader than putting 'em all at the end.

I have been careful to procure the best Paper; that I might in some measure come up to the Beauty of the Latin Edition, tho this bear but half the Price of it.

And I hope the Translator has express'd the Author's Sense aright, and has not committed Faults beyond what an ingenuous Reader can pardon.

By the Corrector's fault *Knowlege* is spelt thro-out the Book without a *d*.

NEW

NEW
CONJECTURES
Concerning the

Planetary Worlds,
THEIR
INHABITANTS
AND
PRODUCTIONS.

Written by CHRISTIANUS HUY-
GENS, and infcrib'd to his Brother
CONSTANTINE HUYGENS.

BOOK the Firſt.

A Man that is of *Copernicus*'s
Opinion, that this Earth of
ours is a Planet, carry'd
round and enlighten'd by the
Sun, like the reſt of them, cannot but
ſome-

Book 1. fometimes have a fancy, that it's not improbable that the reft of the Planets have their Drefs and Furniture, nay and their Inhabitants too as well as this Earth of ours : Efpecially if he confiders the later Difcoveries made fince *Copernicus*'s time of the Attendents of *Jupiter* and *Saturn,* and the Champain and hilly Countrys in the Moon, which are an Argument of a relation and kin between our Earth and them, as well as a proof of the Truth of that Syftem.   This has often been our talk, I remember, good Brother, over a large Telefcope, when we have been viewing thofe Bodies, a ftudy that your continual bufinefs and abfence have interrupted for this many years.   But we were always apt to conclude, that 'twas in vain to enquire after what Nature had been pleafed to do there, feeing there was no likelihood of ever coming to an end of the Enquiry.   Nor could I ever find that any Philofophers, thofe bold Heros, either antient or modern, ventur'd fo far.   At the very birth of Aftronomy, when the Earth was firft afferted to
be

be Spherical, and to be furrounded Book 1.
with Air, even then there were fome
men fo bold as to affirm, there were *Some have already*
an innumerable company of Worlds *talk'd of*
in the Stars. But later Authors, fuch *the Inhabitants of the*
as Cardinal *Cufanus*, *Brunus*, *Kepler*, *Planets*,
(and if we may believe him, *Tycho* was *but went*
of that opinion too) have furnifh'd *no farther.*
the Planets with Inhabitants. Nay,
Cufanus and *Brunus* have allow'd the
Sun and fixed Stars theirs too. But
this was the utmoft of their boldnefs ;
nor has the ingenious French Author
of the Dialogues about *the Plurality
of Worlds* carry'd the bufinefs any farther. Only fome of them have coined
fome pretty Fairy Stories of the Men
in the Moon, juft as probable as *Lu-
cian*'s true Hiftory ; among which I
muft count *Kepler*'s, which he has di-
verted us with in his Aftronomical
Dream. But a while ago thinking
fomewhat ferioufly of this matter (not
that I count my felf quicker fighted
than thofe great Men, but that I had
the happinefs to live after moft of
them) methoughts the enquiry was
not fo impracticable, nor the way fo
ftopt

4   *Conjectures concerning*

ſtopt up with Difficulties, but that there was very good room left for probable Conjectures. As they came into my head, I clapt them down into common places, and ſhall now try to digeſt them into ſome tolerable Method for your better conception of them, and add ſomewhat of the Sun and Fixt Stars, and the Extent of that Univerſe of which our Earth is but an inconſiderable point. I know you have ſuch an eſteem and reverence for any thing that belongs to Heaven, that I perſwade my ſelf you will read what I have written without pain: I'm ſure I writ it with a great deal of pleaſure; but as often before, ſo now, I find the ſaying of *Archytas* true, even to the Letter, *That tho a Man were admitted into Heaven to view the wonderful Fabrick of the World, and the Beauty of the Stars, yet what would otherwiſe be Rapture and Extaſie, would be but a melancholy Amazement if he had not a Friend to communicate it to.* I could wiſh indeed that all the World might not be my Judges, but that I might chuſe my Readers, Men like you, not
ignorant

ignorant in Aftronomy and true Philofophy ; for with fuch I might pro-
mife my felf a favourable hearing, and
not need to make an Apology for da-
ring to vent any thing new to the
World. But becaufe I am aware what
other hands it's likely to fall into, and
what a dreadful Sentence I may expect
from thofe whofe Ignorance or Zeal is
too great, it may be worth the while
to guard my felf beforehand againft
the Affaults of thofe fort of People.

There's one fort who knowing nothing of Geometry or Mathematicks,
will laugh at it as a whimfical and ridiculous undertaking. It's mere Con-
juration to them to talk of meafuring
the Diftance or Magnitude of the
Stars : And for the Motion of the
Earth, they count it, if not a falfe, at
leaft a precarious Opinion; and no
wonder then if they take what's built
upon fuch a flippery Foundation for
the Dreams of a fanciful Head and a
diftemper'd Brain. What fhould we
anfwer to thefe Men, but that their
Ignorance is the caufe of their Diflike,
and that if they had more Senfe they
would

The Obje-
ctions of
ignorant
Cavillers
prevented.

would have fewer Scruples? But few people having had an opportunity of profecuting thefe Studies, either for want of Parts, Learning, or Leifure, we cannot blame their Ignorance; and if they refolve to find fault with us for fpending time in fuch matters, becaufe they do not underftand the ufe of them, we muft appeal to properer Judges.

Thefe Con-jectures do not cnotra-dict the ho-ly Scrip-tures. The other fort, when they hear us talk of new Lands, and Animals endued with as much Reafon as themfelves, will be ready to fly out into religious Exclamations, that we fet up our Conjectures againft the Word of God, and broach Opinions directly oppofite to Holy Writ. For we do not there read one word of the Production of fuch Creatures, no not fo much as of their Exiftence; nay rather we read the quite contrary. For, That only mentions this Earth with its Animals and Plants, and Man the Lord of them; but as for Worlds in the Sky, 'tis wholly filent. Either thefe Men refolve not to underftand, or they are very ignorant; For they have

been

been anfwer'd fo often, that I am almoft afham'd to repeat it : That it's
evident God had no defign to make a
particular Enumeration in the Holy
Scriptures, of all the Works of his
Creation. When therefore it is plain
that under the general name of *Stars*
or *Earth* are comprehended all the
Heavenly Bodies, even the little Gentlemen round *Jupiter* and *Saturn*, why
muft all that multitude of Beings
which the Almighty Creator has been
pleafed to place upon them, be ex-
cluded the Privilege, and not fuffer'd
to have a fhare in the Expreffion?
And thefe Men themfelves can't but
know in what fenfe it is that all things
are faid to be made for the ufe of Man,
not certainly for us to ftare or peep
through a Telefcope at ; for that's little better than nonfenfe. Since then
the greateft part of God's Creation,
that innumerable multitude of Stars, is
plac'd out of the reach of any man's
Eye ; and many of them, it's likely,
of the beft Glaffes, fo that they don't
feem to belong to us ; is it fuch an un-
reafonable Opinion, that there are
fome

Book 1. some reasonable Creatures who see and admire those glorious Bodies at a nearer distance?

This Enquiry not over-curious. But perhaps they'll say, it does not become us to be so curious and inquisitive in these things which the Supreme Creator seems to have kept for his own knowlege: For since he has not been pleased to make any farther Discovery or Revelation of them, it seems little better than presumption to make any inquiry into that which he has thought fit to hide. But these Gentlemen must be told, that they take too much upon themselves when they pretend to appoint how far and no farther Men shall go in their Searches, and to set bounds to other Mens Industry; just as if they had been of the Privy Council of Heaven: as if they knew the Marks that God has plac'd to Knowlege: or as if Men were able to pass those Marks. If our Forefathers had been at this rate scrupulous, we might have been ignorant still of the Magnitude and Figure of the Earth, or of such a place as *America.* The Moon might have shone

the *Planetary Worlds.* 9

ſhone with her own Light for all us, and we might have ſtood up to the ears in Water, like the *Indians* at every Eclipſe: and a hundred other things brought to light by the late Diſcoveries in Aſtronomy had ſtill been unknown to us. For what can a Man imagine more abſtruſe, or leſs likely to be known, than what is now as clear as the Sun? That vigorous Induſtry, and that piercing Wit were given Men to make advances in the ſearch of Nature, and there's no reaſon to put any ſtop to ſuch Enquiries. I muſt acknowlege ſtill that what I here intend to treat of is not of that nature as to admit of a certain knowlege; I can't pretend to aſſert any thing as poſitively true (for that would be madneſs) but only to advance a probable gueſs, the truth of which every one is at his own liberty to examine. If any one therefore ſhall gravely tell me, that I have ſpent my time idly in a vain and fruitleſs enquiry after what by my own acknowlegement I can never come to be ſure of; the anſwer is, that at this rate he would put down all

Natu-

Book 1. Natural Philofophy as far as it con-
cerns it felf in fearching into the Nature of things: In fuch noble and
Conjectures fublime Studies as thefe, 'tis a Glory
not ufelefs, to arrive at Probability, and the fearch
becaufe not it felf rewards the pains. But there
certain. it felf rewards the pains. But there
are many degrees of Probable, fome
nearer Truth than others, in the deter-
mining of which lies the chief exercife of our Judgment. But befides the
Thefe Stu- Noblenefs and Pleafure of the Studies,
dies ufeful may not we be fo bold as to fay, they
to Religion. are no fmall help to the advancement
of Wifdom and Morality? fo far are
they from being of no ufe at all. For
here we may mount from this dull
Earth, and viewing it from on high,
confider whether Nature has laid out
all her coft and finery upon this fmall
fpeck of Dirt. So, like Travellers into other diftant Countrys, we fhall be
better able to judg of what's done at
home, know how to make a true
eftimate of, and fet its own value upon every thing. We fhall be lefs apt
to admire what this World calls great,
fhall nobly defpife thofe Trifles the
generality of Men fet their Affections
on,

on, when we know that there are a Book 1.
multitude of fuch Earths inhabited and
adorned as well as our own. And we
fhall worfhip and reverence that God
the Maker of all thefe things; we fhall
admire and adore his Providence and
wonderful Wifdom which is difplayed
and manifefted all over the Univerfe,
to the confufion of thofe who would
have the Earth and all things formed
by the fhuffling Concourfe of Atoms,
or to be without beginning. But to
come to our purpofe.

And now becaufe the chief Argu-
ment for the proof of what we intend Coperni-
will be taken from the difpofition of cus's Sy-
the Planets, among which without *stem ex-*
plain'd.
doubt the Earth muft be counted in
the Copernican Syftem, I fhall here
firft of all draw two Figures. The
firft is a Defcription of the Orbs the
Planets move in, in that order that
they are placed round the Sun, drawn
as near as can be in their true Proportions, like what you have feen in my
Clock at home. The fecond fhows
the Proportions of their Magnitudes
in refpect of one another and of the
Sun,

Book 1. Sun, which you know is upon that
same Clock of mine too. In the first
the middle Point or Center is the Place
of the Sun, round which, in an order
that every one knows, are the Orbits of
Mercury, Venus, the Earth with that
of the Moon about it ; then those of
Mars, Jupiter and *Saturn :* and about
the two last the small Circles that
their Attendents march in : about *Ju-*
piter four, and about *Saturn* five.
Which Circles as well as that of the
Moon are drawn larger than their
true Proportion would admit, otherwise they could not have been seen.
You may easily apprehend the Vast-
ness of these Orbits by this, that the
distance of the Earth from the Sun is
ten or twelve thousand of the Earth's
Diameters. Almost all these Circles
are in the same Plane, declining very
little from that in which the Earth
moves, call'd *the Plane of the Eclip-*
tick. This Plane is cut obliquely by
the Axis upon which the Earth turns it
self round in 24 hours, whence arise
the Successions of Day and Night :
The Axis of the Earth always keeping

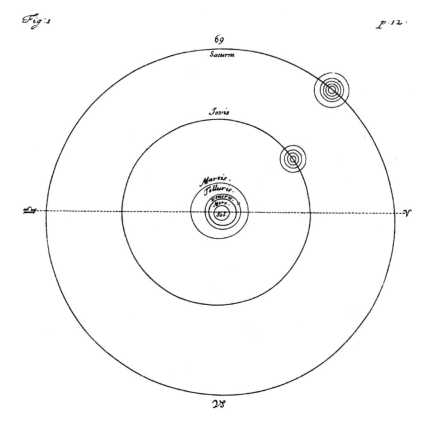

ing the fame Inclination to the Ecliptick (except a fmall change beft ∿,
known to Aftronomers) while the
Earth it felf is carry'd in its yearly
Courfe round the Sun, caufes the regular Order of the Seafons of the Year:
as you may fee in all Aftronomers
Books. Out of which I fhall tran-
fcribe hither the Periods of the Revolutions of the Planets, *viz.* *Saturn*
moves round the Sun in 29 Years, 174
Days, and 5 Hours: *Jupiter* finifhes
his Courfe in 11 Years, 317 Days,
and 15 Hours: *Mars* his in about 687
Days. Our Year is 365 Days 6 Hours:
Venus's 224 Days 18 Hours: and
Mercury's 88 Days. This is the now
commonly receiv'd Syftem, invented
by *Copernicus*, and very agreeable to
that frugal Simplicity Nature fhows in *Arguments*
all her Works. If any one is refolved *for the*
to find fault with it, let him firft be *truth of it.*
fure he underftands it. Let him firft
fee in the Books of Aftronomers with
how much greater eafe and plainnefs
all the Motions of the Stars, and Appearances in the Heavens are explained
and demonftrated in this than either in
that

Book 1. that of *Ptolomy* or *Tycho*. Let him
consider that Discovery of *Kepler*, that
the distances of the Planets from the
Sun, as well of the Earth as the rest,
are in a fixt certain proportion to the
times they spend in their Revolutions.
Which Proportion it's since observed
that their Satellites keep round *Jupiter*
and *Saturn*. Let him examine what
a contradictory Motion they are fain
to invent for the solution of the Polar
Star's changing its distance from the
Pole. For that Star in the end of the
Little Bear's Tail which now describes
so small a Circle round the Pole, that
it is not above two Degrees and twenty Minutes, was observed about 1820
Years ago, in the time of *Hipparchus*,
to be above 12 : and will within a
few ages more be forty five Degrees
distant from it : and after 25000 years
more will return to the same place it
is now in. Now if with them we allow the Heavens to be turned upon
their own Axis, at this rate they must
have a new Axis every day : a thing
most abominably absurd, and repug-
nant to the nature of all motion. Whereas

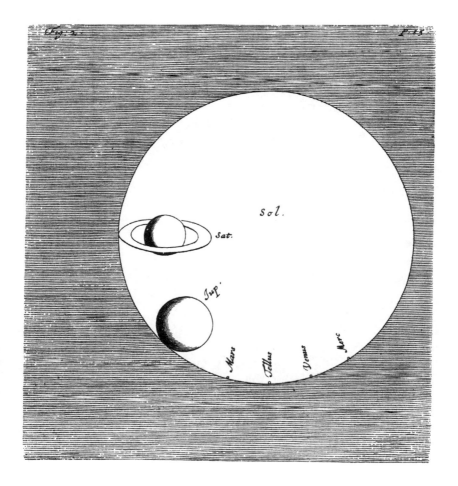

Fig. 2.

as nothing is eafier with *Copernicus* Book 1.
than to give us fatisfaction in this matter. Then he may impartially weigh
thofe Anfwers that *Galilæus, Gaffen-
dus, Kepler,* and others have given to
all Objections propofed, which have fo
fatisfied all Scruples, that generally all
Aftronomers now adays are brought
over to our fide, and allow the Earth
its Motion and Place among the Planets. If he cannot be fatisfied with
all this, he is either one whofe Dulnefs
can't comprehend it, or who has his
Faith at another man's difpofal, and
fo for fear of *Galileo*'s fate dare not
own it.

In the other Figure you have the
Globes of the Planets, and of the Sun,
reprefented to your eyes as plac'd near
one another. Where I have obferv'd *The Pro-*
the fame Proportion, of their Diame- *portion of*
ters to that of the Sun, that I pub- *the Mag-*
lifh'd to the World in my Book of *the Pla-*
the *Appearances of Saturn:* namely, *nets, in re-*
the Diameter of the Ring round *Sa-* *another,*
turn is to that of the Sun as 11 is to *and the*
37; that of *Saturn* himfelf about as 5 *Sun.*
to 37; that of *Jupiter* as 2 to 11; that
of

Book 1. of *Mars* as 1 to 166 ; of the Earth as
1 to 111 ; and of *Venus* as 1 to 84 : to
which I shall now add that of *Mercu-*
ry observ'd by *Hevelius* in the Year
1661, but calculated by my self, and
found to be as 1 to 290.

If you would know the way that
we came to this knowledg of their
Magnitudes, by knowing the Proportion of their Distances from the Sun,
and the measure of their Diameters,
you may find it in the Book before-
mentioned : and I cannot yet see any
reason to make an alteration in those I
then settled, altho I will not say they
are without their faults.   For I can't

The La- yet be of their mind, who think the
mellæ more
convenient use of Micrometers, as they call them,
than Mi- is beyond that of our Plates, but must
crometers. still think that those thin Plates or Rods
of which I there taught the use, not to
detract from the due praises of so use-
ful an Invention, are more convenient
than the Micrometers.

In this Proportion of the Planets it
is worth while to take notice of the
prodigious Magnitude of the Sun in
comparison with the four innermost,
which

which are far lefs than *Jupiter* and *Sa-* Book I
turn.   And 'tis remarkable, that the
Bodies of the Planets do not increafe
together with their diftances from the
Sun, but that *Venus* is much bigger
than *Mars*.

Having thus explain'd the two
Schemes, there's no body I fuppofe
but fees, that in the firft the Earth is *The Earth*
made to be of the fame fort with the *juftly li-*
reft of the Planets.   For the very Po- *ken'd to*
fition of the Circles fhows it.   And *nets, and*
that the other Planets are round like it, *the Pla-*
and like it receive all the Light they *nets to it.*
have from the Sun, there's no room
(fince the Difcoveries made by Telefcopes) to doubt.   Another thing they
are like it in is, that they are moved
round their own Axis ; for fince 'tis
certain that *Jupiter* and *Saturn* are,
who can doubt it of the others? Again,
as the Earth has its Moon moving
round it, fo *Jupiter* and *Saturn* have
theirs.   Now fince in fo many things
they thus agree, what can be more
probable than that in others they agree
too ; and that the other Planets are
as beautiful and as well ftock'd with
In-

Book 1. Inhabitants as the Earth? or what shadow of Reason can there be why they should not?

If any one should be at the diffecti-on of a Dog, and be there shewn the Intrails, the Heart, Stomach, Liver, Lungs and Guts, all the Veins, Arte-ries and Nerves; could such a Man reasonably doubt whether there were the same Contexture and Variety of Parts in a Bullock, Hog, or any other Beast, tho he had never chanc'd to see the like opening of them? I don't be-lieve he would. Or were we tho-rowly satisfy'd in the Nature of one of the Moons round *Jupiter*, should not we straight conclude the same of the rest of them? So if we could be assur'd in but one Comet, what it was that is the cause of that strange ap-pearance, should we not make that a Standard to judg of all others by?

Arguments from their Similitude of no small weight. 'Tis therefore an Argument of no small weight that is fetch'd from Rela-tion and Likeness; and to reason from what we see and are sure of, to what we cannot, is no false Logick. This must be our Method in this Treatise, where-

wherein from the Nature and Circum-
ftances of that Planet which we fee
before our eyes, we may guefs at thofe
that are farther diftant from us.

And, Firft, 'tis more than probable *The Pla-*
that the Bodies of the Planets are fo- *nets are*
lid like that of our Earth, and that *not with-*
folid, and
they don't want what we call Gravi- *out Gravi-*
ty, that Virtue, which like a Load- *ty.*
ftone attracts whatfoever is near the
Body to its Center. And that they
have fuch a Quality, their very Figure
is a proof ; for their Roundnefs proceeds only from an equal preffure of
all their Parts tending to the fame Center. Nay more, we are fo skilful
now adays, as to be able to tell how
much more or lefs the Gravitation in
Jupiter or *Saturn* is than here ; of
which Difcovery and its Author you
may read my *Effay of the Caufes of Gra-*
vitation.

But now to carry the fearch farther,
let us fee by what fteps we muft rife to
the attaining fome knowlege in the
more private Secrets concerning the
State and Furniture of thefe new
Earths. And, firft, how likely is it
                                        that

that they may be ſtock'd with Plants
and Animals as well as we? I ſuppoſe
no body will deny but that there's
ſomewhat more of Contrivance, ſomewhat more of Miracle in the production and growth of Plants and Animals,
than in lifeleſs heaps of inanimate Bodies, be they never ſo much larger;
as Mountains, Rocks, or Seas are.
For the finger of God, and the Wiſdom of Divine Providence, is in them
much more clearly manifeſted than in
the other. One of *Democritus*'s or
Cartes's Scholars may venture perhaps
to give ſome tolerable Explication of
the appearances in Heaven and Earth,
allow him but his Atoms and Motion;
but when he comes to Plants and Animals, he'll find himſelf non-plus'd,
and give you no likely account of their
Production. For every thing in them
is ſo exactly adapted to ſome deſign,
every part of them ſo fitted to its proper uſe, that they manifeſt an Infinite
Wiſdom, and exquiſite Knowlege in
the Laws of Nature and Geometry,
as, to omit thoſe Wonders in Generation, we ſhall by and by ſhow; and
make

make it an abſurdity even to think of their being thus haply jumbled toge- ther by a chance Motion of I don't know what little Particles. Now ſhould we allow the Planets nothing but vaſt Deſerts, lifeleſs and inanimate Stocks and Stones, and deprive them of all thoſe Creatures that more plain- ly ſpeak their Divine Architect, we ſhould ſink them below the Earth in Beauty and Dignity; a thing that no Reaſon will permit, as I ſaid before.

Well then, now we have gain'd the Point for them, and the Planets may be allow'd ſome Bodys capable of moving themſelves, not at all inferior to ours, (for why ſhould they?) and theſe are Animals. Now for fear of ſtarving theſe poor Creatures, we muſt have Plants you know. And ſo the other Point is gain'd. And as for their Growth and Nouriſhment, 'tis no *Not to be* doubt the ſame with ours, ſeeing they *imagin'd too unlike* have the ſame Sun to warm and en- *ours.* liven them as ours have.

But perhaps ſome body may ſay, we conclude too faſt. They will not deny indeed but that there may be
Plants

Book 1. Plants and Animals on the Surface of
the Planets, that deferve as well to be
provided for by their Creator as ours
do : but why muft they be of the fame
nature with ours? Nature feems to
court variety in her Works, and may
have made them widely different from
ours either in their matter or manner
of Growth, in their outward Shape,
or their inward Contexture ; fhe may
have made them fuch as neither our
Underftanding nor Imagination can
conceive. That's the thing we fhall
now examin, and whether it be not
more likely that fhe has not obferv'd
fuch a variety as they talk of. Nature
feems moft commonly, and in moft of
her Works, to affect Variety, 'tis true ;
But they fhould confider 'tis not the
bufinefs of a man to pretend to fettle
how great this Difference and Variety
muft be. Nor does it follow, becaufe
it may be Infinite, and out of our comprehenfion and reach, that therefore
things in reality are fo. For fuppofe
God fhould have pleafed to have made
all things there juft as he has here, the
Inhabitants of thofe Places (if there
are

are any fuch ftrange things) would
admire his Wifdom and Contrivance
no lefs than if they were widely different; feeing they can't come to know
what's done in the other Planets.
Who doubts but that God, if he had
pleafed, might have made the Animals in *America* and other diftant
Countries nothing like ours? (and Nature you know affects Variety) yet
we fee he has not done it. They have
indeed fome difference in their fhape,
and 'tis fit they fhould, to diftinguifh
the Plants and Animals of thofe Countries from ours, who live on this fide
the Earth; but even in this variety
there is an Agreement, an exact Cor-
refpondence in figure and fhape, the
fame ways of Growth, and new Productions, and of continuing their own
kind. Their Animals have Feet and
Wings like ours, and like ours have
Heart, Lungs, Guts, and the Parts
ferving to Generation; whereas all
thefe things, as well with them as us,
might, if it had fo pleafed Infinite Wifdom, have been order'd a very different way. 'Tis plain then that Nature

ture has not exhibited that Variety in her Works that she could, and therefore we muft not allow that weight to this Argument, as upon the account of it to make every thing in the Planets quite different from what is here. 'Tis more probable that all the difference there is between us and them, fprings from the greater or lefs diftance and influence from that Fountain of Heat and Life the Sun ; which will caufe a difference not fo much in their Form and Shape, as in their Matter and Contexture.

Planets have Water. And as for the matter whereof the Plants and Animals there confift, tho it is impoffible ever to come to the knowlege of its Nature, yet this we may venture to affert (there being fcarce any doubt of it) that their Growth and Nourifhment proceeds from fome liquid Principle. For all Philofophers agree that there can be no other way of Nutrition ; fome of the chief among them having made Water to be the Original of all things : For whatfoever's dry and without moifture, is without motion too ; and

with-

without motion it's impoſſible there Book 1.
ſhould be any increaſe. But the parts ∿
of a Liquid being in continual motion
one with another, and infinuating and
twifting themſelves into the ſmalleſt
Places, are thereby very proper and
apt to add not themſelves only, but
whatſoever elſe they may bring along
with them to the increaſe and growth
of Bodies. Thus we ſee that by the
means of Water the Plants grow, bloſ-
ſom, and bear Fruit ; and by the addition of that only, Stones grow together out of Sand. And there's no
doubt but that Metals, Cryftals, and
Jewels, have the ſame method of
Production: Tho in them there has
been no opportunity to make the ſame
obſervation, as well by reaſon of their
ſlow advances, as that they are commonly found far from the Places of
their Generation ; thrown up I ſuppoſe by ſome Earthquakes or Convul-
ſions. That the Planets are not without Water, is made not improbable
by the late Obſervations: For about
Jupiter are obſerv'd ſome ſpots of a
darker hue than the reſt of his Body,
which

Book 1. which by their continual change show
themselves to be Clouds: For the spots
of *Jupiter* which belong to him, and
never remove from him, are quite
different from these, being sometimes
for a long time not to be seen for these
Clouds; and again, when these dis-
appear, showing themselves. And at
the going off of these Clouds, some
spots have been taken notice of in him,
much brighter than the rest of his Body, which remain'd but a little while,
and then were hid from our sight.
These Monsieur *Cassini* thinks are only the Reflection from the Snow that
covers the tops of the Hills in *Jupiter :*
but I should rather think that it is only the colour of the Earth, which
chances to be free from those Clouds
that commonly darken it.

Mars too is found not to be without
his dark spots, by means of which he
has been observ'd to turn round his
own Axis in 24 hours and 40 minutes;
the length of his day : but whether he
has Clouds or no, we have not had
the same opportunity of observing as
in *Jupiter*, as well because even when
he

he is neareſt the Earth, he appears to Book 1.
us much leſs than *Jupiter*, as that his
Light not coming ſo long a Journey,
is ſo brisk as to be an Impediment to
exaƐt Obſervations : And this Reaſon
is as much ſtronger in *Venus* as its Light
is. But ſince 'tis certain that the Earth
and *Jupiter* have their Water and
Clouds, there is no reaſon why the
other Planets ſhould be without them.
I can't ſay that they are exaƐtly of the *But not*
ſame nature with our Water; but that *juſt like*
they ſhould be liquid their uſe requires, *ours.*
as their beauty does that they ſhould
be clear. For this Water of ours, in
Jupiter or *Saturn*, would be frozen
up inſtantly by reaſon of the vaſt diſtance of the Sun. Every Planet therefore muſt have its Waters of ſuch a
temper, as to be proportion'd to its
heat : *Jupiter*'s and *Saturn*'s muſt be
of ſuch a nature as not to be liable to
Froſt; and *Venus*'s and *Mercury*'s of
ſuch, as not to be eaſily evaporated by
the Sun. But in all of them, for a
continual ſupply of Moiſture, whatever Water is drawn up by the Heat
of the Sun into Vapors, muſt neceſſarily

rily return back again thither. And this it cannot do but in drops, which are caufed as well there as with us, by their afcending into a higher and colder Region of the Air, out of that which, by reafon of the Reflection of the Rays of the Sun from the Earth, is warmer and more temperate.

Here then we have found in thefe new Worlds Fields warm'd by the kindly Heat of the Sun, and water'd with fruitful Dews and Showers: That there muft be Plants in them as well for Ornament as Ufe, we have fhewn juft now. And what Nourifh-ment, what manner of Growth fhall *Plants* we allow them? Why, I think there *grow and* can be no better, nay no other, than *are nou-* *rifh'd there* what we here experience; by having *as they are* their Roots faftned into the Earth, and *here.* imbibing its nourifhing Juices by their tender Fibres. And left they fhould be only like fo many bare Heaths, with nothing but creeping Shrubs and Bufhes, we'll e'en fend them fome no-bler and loftier Plants, Trees, or fome-what like them: Thefe being the greateft, and, except Waters, the only

Orna-

Ornament that Nature has beſtow'd Book 1. upon the Earth. For not to ſpeak of thoſe many uſes that are made of their Wood, there's no one that is ignorant either of their Beauty or Pleaſantneſs. Now what way can any one imagine for a continual Production and Succeſſion of theſe Plants, but their bearing Seed? A Method ſo excellent that it's the only one that Nature has here made uſe of, and ſo wonderful, that it ſeems to be deſign'd not for this Earth alone. In fine, there's the ſame reaſon to think that this Method is obſerv'd in thoſe diſtant Countries, as there was of its being follow'd in the remote Quarters of this ſame Earth.

'Tis much the ſame in Animals as *The ſame* 'tis in Plants, as to their manner of *true of* Nouriſhment, and Propagation of their *their Ani-* kind. For ſince all the living Crea- *mals.* tures of this Earth, whether Beaſts, Birds, Fiſhes, Worms, or Inſects, u- niverſally and inviolably follow the ſame conſtant and fixt Inſtitution of Nature; all feed on Herbs, or Fruits, or the Fleſh of other Animals that fed on

on them : fince all Generation is perform'd by the impregnating of the
Eggs, and the Copulation of Male and
Female: Why may not the fame
rule be obferv'd in the Planetary
Worlds? For *'tis certain that the Herbs
and Animals that are there would be loft,
their whole Species deftroy'd without fome
daily new Productions:* except there be
no fuch thing there as Misfortune or
Accident: except the Plants are not
like other humid Bodies, but can bear
Heat, Froft and Age, without being
dry'd up, kill'd, or decay'd : except
the Animals have Bodies as hard and
durable as Marble ; which I think are
grofs Abfurdities. If we fhould invent fome new way for their coming
into the World, and make them drop
like Soland Geefe from Trees, how ridiculous would this be to any one that
confiders the vaft difference between
Wood and Flefh? Or fuppofe we
fhould have new ones made every day
out of fome fuch fruitful Mud as that
of *Nile,* who does not fee how contrary this is to all that's reafonable?
And that 'tis much more agreeable to
the

the Wifdom of God, once for all to create of all forts of Animals, and diftribute them all over the Earth in fuch a wonderful and inconceivable way as he has, than to be continually obliged to new Productions out of the Earth? And what miferable, what helplefs Creatures muft thefe be, when there's no one that by his duty will be obliged, or by that ftrange natural fondnefs, which God has wifely made a neceffary argument for all Animals to take care of their own, will be moved to affift, nurfe or educate them?

As for what I have faid concerning their Propagation, I cannot be fo pofitive; but the other thing, namely, that they have Plants and Animals, I think I have fully proved. And by the fame Argument, of their not being inferiour to our Earth, they muft have as great a variety of both as we have. What this is, will be beft known to him that confiders the different ways our Animals make ufe of in moving from one place to another. Which may be reduc'd, I think, to thefe; either

ther

ther that they walk upon two feet or four ; or like Infects, upon fix, nay fometimes hundreds; or that they fly in the Air bearing up, and wonderfully fteering themfelves with their Wings ; or creep upon the Ground without feet ; or by a violent Spring in their Bodies, or paddling with their feet, cut themfelves away in the Waters. I don't believe, nor can I conceive, that there fhould be any other way than thefe mention'd. The Animals then in the Planets muft make ufe of one or more of thefe, like our amphibious Birds, which can fwim in Water as well as walk on Land, or fly in the Air ; or like our Crocodiles and Sea-Horfes, muft be Mongrels, between Land and Water. There can no other method be imagin'd but one of thefe. For where is it poffible for Animals to live, except upon fuch a folid Body as our Earth, or a fluid one like the Water, or ftill a more fluid one than that, fuch as our Air is? The Air I confefs may be much thicker and heavier than ours, and fo, without any difadvantage to its Tranfparency,

rency, be fitter for the volatile Animals. There may be too many forts of Fluids ranged over one another in rows as it were. The Sea perhaps may have fuch a fluid lying on it, which tho ten times lighter than Water, may be a hundred times heavier than Air ; whofe utmoft Extent may not be fo large as to cover the higher places of their Earth. But there's no reafon to fufpect or allow them this, fince we have no fuch thing ; and if we did, it would be of no advantage to them, for that the former ways of moving would not be hereby at all increas'd : But when we come to meddle with the Shape of thefe Creatures, and confider the incredible variety that is even in thofe of the different parts of this Earth, and that *America* has fome which are no where elfe to be found, I muft then confefs that I think it beyond the force of Imagination to arrive at any knowlege in the matter, or reach probability concerning the figures of thefe Planetary Animals. Altho confidering thefe ways of Motion we e'en now recounted, they

they may perhaps be no more different from ours than ours (thoſe of ours I mean that are moſt unlike) are from one another.

If a man were admitted to a Survey of *Jupiter* or *Venus*, he would no doubt find as great a number and variety as he had at home. Let us then, that we may make as near a gueſs at, and as reaſonable a judgment of the matter as we can, conſider the many ſorts, and the admirable difference in *Great va-* the ſhapes of our own Animals ; run-
riety of A-
nimals in ning over ſome of the chief of them
this Earth. (for 'twould be tedious to ſet about a general Catalogue) that are notoriouſly different from one another, either in their Figure or ſome peculiar Property belonging to them ; as they belong to the Land, or the Water, or the Air. Among the Beaſts we may take notice of the great diſtance between the Horſe, the Elephant, the Lion, the Stag, the Camel, the Hog, the Ape, the Porcupine, the Tortoiſe, the Cameleon : in the Water, of that between the Whale, and the Sea-Calf, the Skait, the Pike, the Eel, the Ink-
Fiſh,

Fifh, the Pourcontrel, the Crocodile, Book 1.
the flying Fifh the Cramp Fifh, the 〰
Crab, the Oifter, and the Purple Fifh :
and among Birds, of that between the
Eagle, the Oftrich, the Peacock, the
Swan, the Owl, and the Bat : and in
Infects, of that between the Ants, the
Spider, the Fly, and the Butterfly ;
and of that Prodigy in their wonderful change from Worms. In this Roll
I have pafs'd by the creeping kind as
one fort, and skip'd over that vaft
multitude of lefs different Animals
that fill the intermediate fpaces. But
be they never fo many, there is no
reafon to think that the Planets cannot *And no lefs*
match them. For tho we in vain guefs *in the Pla-*
at the Figures of thofe Creatures, yet *nets.*
we have difcover'd fomewhat of their
manner of Life in general ; and of their
Senfes we fhall more by and by.

The more confiderable Differences *The fame*
in our Plants ought to be thought on, *in Plants.*
as well as the other. As in Trees,
that between the Fir and the Oak, the
Palm, the Vine, the Fig, and the Co-
co-Nut Tree, and that in the *Indies,*
from whofe Boughs new Roots fpring,
and

Book 1. and grow downwards into the Earth.
In Herbs, the difference is notable between Grafs, Poppy, Colewort, Ivy,
Pompions, and the Indian Fig with
thick Leaves growing up without any
Stalk, and Aloe. Between every one
of which again there are many lefs
differing Plants not taken notice of.
Then the different ways of raifing
them are remarkable, whether from
Seeds, or Kernels, or Roots, or by
grafting or inoculating them. And
yet in all thefe, whether we confider
the things themfelves, or the ways of
their Production, I make no doubt
but that the Planetary Worlds have as
wonderful a variety as we.

Rational   But ftill the main and moft divert-
Animals   ing Point of the Enquiry is behind,
in the Pla- which is the placing fome Spectators
nets.   in thefe new Difcoveries, to enjoy
thefe Creatures we have planted them
with, and to admire their Beauty and
Variety. And among all, that have
never fo flightly meddled with thefe
matters, I don't find any that have
fcrupled to allow them their Inhabitants : not Men perhaps like ours, but
fome

some Creatures or other endued with Reason. For all this Furniture and Beauty the Planets are stock'd with seem to have been made in vain, without any design or end, unless there were some in them that might at the same time enjoy the Fruits, and adore the wise Creator of them. But this alone would be no prevailing Argument with me to allow them such Creatures. For what if we should say, that God made them for no other design, but that he himself might see (not as we do 'tis true; but that he that made the Eye sees, who can doubt?) and delight himself in the contemplation of them? For was not Man himself, and all that the whole World contains, made upon this very account? That which makes me of this opinion, that those Worlds are not without such a Creature endued with Reason, is, that otherwise our Earth would have too much the advantage of them, in being the only part of the Universe that could boast of such a Creature so far above, not only Plants and Trees, but all

Ani-

Book 1. Animals whatfoever : a Creature that
has a Divine fomewhat within
him, that knows, and underftands,
and remembers fuch an innumerable
number of things ; that deliberates,
weighs and judges of the Truth : a
Creature upon whofe account, and for
whofe ufe, whatfoever the Earth
brings forth feems to be provided. For
every thing here he converts to his
own ends. With the Trees, Stones,
and Metals, he builds himfelf Houfes :
the Birds and Fifhes he fuftains himfelf with : and the Water and Winds
he makes fubfervient to his Navigation ; as he doth the fweet Smell and
glorious Colours of the Flowers to his
Delight. What can there be in the
Planets that can make up for its De-
fects in the want of fo noble an Animal ? If we fhould allow *Jupiter* a
greater variety of other Creatures,
more Trees, Herbs and Metals, all
thefe would not advantage or dignify
that Planet fo much as that one Animal doth ours by the admirable Productions of his penetrating Wit. If I
am out in this, I do not know when
to

to truſt my Reaſon, and muſt allow Book 1.
my ſelf to be but a poor Judg in the
true eſtimate of things.

Nor let any one ſay here, that there's *Vices of*
ſo much Villany and Wickedneſs in *Men no hindrance*
this Man that we have thus magnified, *to their be-*
that it's a reaſonable doubt, whether *ing the Glo-*
he would not be ſo far from being the *ry of the Planet*
Glory and Ornament of the Planet *they inha-*
that enjoys his Company, that he *bit.*
would be rather its Shame and Diſ-
grace. For firſt, the Vices that moſt
Men are tainted with, are no hindrance, but that thoſe that follow the
Dictates of true Reaſon, and obey the
Rules of a rigid Virtue, are ſtill a
Beauty and Ornament to the place that
has the happineſs to harbour them.
Beſides, the Vices of Men themſelves are
of excellent uſe, and are not permitted
and allow'd in the World without
wiſe deſign. For ſince it has ſo pleaſed
God to order the Earth, and every
thing in it as we ſee it is (for it's non-
ſenſe to ſay it happen'd againſt his
Will or Knowlege) we muſt not
think that thoſe different Opinions, and
that various multiplicity of Minds
were

Book 1. were plac'd in different Men to no end
or purpofe: but that this mixture of
bad Men with good, and the Confe-
quents of fuch a mixture, as Misfor-
tunes, Wars, Afflictions, Poverty, and
the like, were given us for this very
good end, *viz.* the exercifing our Wits,
and fharpening our Inventions; by
forcing us to provide for our own neceffary defence againft our Enemies.
'Tis to the fear of Poverty and Mifery
that we are beholden for all our Arts,
and for that natural Knowlege which
was the product of laborious Induftry;
and which makes us that we cannot
but admire the Power and Wifdom of
the Creator, which otherwife we
might have pafs'd by with the fame
indifference as Beafts. And if Men
were to lead their whole Lives in an
undifturb'd continual Peace, in no fear
of Poverty, no danger of War, I
don't doubt they would live little better than Brutes, without all knowlege
or enjoyment of thofe Advantages
that make our Lives pafs on with pleafure and profit. We fhould want the
wonderful Art of Writing, if its great
ufe

ufe and neceſſity in Commerce and Book 1.
War had not forc'd out the Invention. 'Tis to thefe we owe our Art of
Sailing, our Art of Sowing, and moſt
of thofe Difcoveries of which we are
Maſters ; and almoſt all the Secrets
in experimental Knowlege. So that
thofe very things that make up their
Indictment againſt Reaſon, are no
ſmall helps to its advancement and
perfection. For thofe Virtues themfelves, Fortitude and Conſtancy, would
be of no ufe if there were no Dangers,
no Adverſity, no Afflictions for their
exercife and trial.

If we ſhould therefore imagine in
the Planets fome fuch reafonable Animal as Man is, adorn'd with the fame
Virtues, and infected with the fame
Vices, it would be fo far from degra-
ding or vilifying them, that while
they want fuch a one, I muſt think
them inferior to our Earth.

Well, but allowing thefe Planetari- *Reaſon*
ans fome fort of Reaſon, muſt it needs *there not*
be the fame with ours? Why truly I *different*
think 'tis, and muſt be fo ; whether *from what*
we confider it as applied to Juſtice and *'tis here.*

Mora-

Book 1. Morality, or exercifed in the Principles and Foundations of Sciene. For
Reafon with us is that which gives us
a true fenfe of Juftice and Honefty,
Praife, Kindnefs and Gratitude: 'tis
that that teaches us to diftinguifh uni-
verfally between Good and Bad ; and
renders us capable of Knowlege and
Experience in it. And can there be
any where a Reafon contrary to this?
or can what we call juft and generous
in *Jupiter* or *Mars* be thought unjuft
Villany ? This is not at all, I don't fay
probable, but poffible. For the aim
and defign of the Creator is every
where the prefervation and fafety of
his Creatures. Now when fuch a
Reafon as we are mafters of, is neceffary for the prefervation of Life, and
promoting of Society (a thing that
they be not without, as we fhall fhow)
would it not be ftrange that the Planetarians fhould have fuch a perverfe fort
of Reafon given them, as would neceffarily deftroy and confound what it
was defign'd to maintain and defend ?
But allowing Morality and Paffions
with thofe Gentlemen to be fomewhat
diffe-

different from ours, and fuppofing they
may act by other principles in what
belongs to Friendfhip, and Anger, Hatred, Honefty, Modefty, and Come-
linefs, yet ftill there would be no
doubt, but that in the fearch after
Truth, in judging of the Confequences
of things, in reafoning, particularly
in that fort which belongs to Magnitude or Quantity, about which their
Geometry (if they have fuch a thing)
is employ'd, there would be no doubt
I fay, but that their Reafon here muft
be exactly the fame, and go the fame
way to work with ours, and that
what's true in one part will hold true
over the whole Univerfe; fo that all
the difference muft lie in the degrees of
Knowlege, which will be proportio-
nal to the Genius and Capacity of the
Inhabitants.

But I perceive I am got a little too *They have*
far: For till I have furnifhed them *Senfes.*
with Senfes, neither will Life be any
pleafure to them, nor Reafon of any
ufe. And I think it very probable,
that all their Animals, as well their
Beafts as rational Creatures, are like
ours

ours in all that relates to the Senses : For without the power of Seeing we fhould find it impoffible for Animals to provide Food for themfelves, or be forewarn'd of any approaching danger, fo as to guard themfelves from it. So that where-ever we plant any Animals, except we would have them lead the Life of Worms or Moles, we muft allow them Sight ; than which nothing can conduce more either to the prefervation or pleafure of their Lives. Then if we confider the wonderful nature of Light, and the amazing Artifice in the fit framing the eye for the reception of it, we cannot but fee that Bodies fo vaftly remote could not be view'd by us in their proper Figures and juft Diftances, any other way than by Sight. For this Senfe, and all others that we know of, muft proceed from an external Motion. Which in the fenfe of Seeing muft come either from the Sun, the fixt Stars, or Fire : whofe Particles being whirled about with a rapid Motion, communicate it to the Celeftial Matter about, whence 'tis convey'd in an inftant to the moft

Sight.

diftant

diftant parts, juft like Sound through
the Air. If it were not for this Motion of the intermediate Matter, we
fhould be all in darknefs, and have
fight neither of Sun nor Stars, nor any
thing elfe, for all other Light muft
come to us at fecond-hand from them.
This Motion perceived by the Eyes is
called Light. And the nice Curiofity
of this Perception is admirable, in
that it is caufed by the fmalleft Parti-
cle of that fine Matter, and can at the
fame time determine the Coaft from
whence the Motion comes ; in that all
thefe different Roads of Motion, thefe
Waves croffing and interfering with
one another, are yet no hindrance to
every ones free paffage. All thefe
things are fo wifely, fo wonderfully
contrived, that it's above the power of
humane Wit, not to invent or frame
fomewhat like them, but even to imagine and comprehend them. For what
can be more amazing, than that a Par-
ticle of Body fhould be fo devifed and
framed, as by its means to fhow us
the Shape, the Pofition, the Diftance,
and all the Motions, nay and all the
Colours,

Book 1. Colours, diftinguifhing of a Body that
is far remote from us? And then the
artful Compofition of the Eye, drawing an exact Picture of the Objects
without it, upon the concave fide of
the Choroides, is even above all admiration, nor is there any thing in which
God has more plainly manifefted his
excellent Geometry. And thefe things
are not only contrived and framed
with fo great Wifdom and Skill, as
not to admit of better, but to any one
that confiders them attentively, they
feem to be of fuch a nature as not to
allow any other Method. For it's
impoffible that Light fhould reprefent
Objects to us at fo vaft a diftance, except by fuch an intervening Motion;
and it's as impoffible that any other
Compofition of the Eye fhould be e-
qually fitted to the reception of fuch
Impreffions. So that I cannot but
think them mightily out, that maintain thefe things might have been contrived many other ways. It's likely
then, and credible, that in thefe things
the Planets have an exact correfpon-
dence with us, and that their Animals
have

have the fame Organs, and ufe the
fame way of fight that we do. Well
then they have Eyes, and two at leaft
we muft grant them, otherwife they
would not perceive fome things clofe
to them, and fo could not avoid Mif-
chiefs that take them on the blind fide.
And if we muft allow them all Animals for the prefervation of their Life,
how much more muft they that make
more, and more noble ufes of them, not
be deprived of the Bleffing of fo advantageous Members? For by them
we view the various Flowers, and the
elegant Features of Beauty : with
them we read, we write, we contem-
plate the Heavens and Stars, and meafure their Diftances, Magnitudes, and
Journeys : which how far they are
common to the Inhabitants of thofe
Worlds with us, I fhall ftrait examine.
But firft I fhall enquire whether now
we have given them one, we may not
venture upon the other four Senfes, to
make them as good Men as our felves.
And truly Hearing puts in hard, and *Hearing.*
almoft perfwades me to give it a fhare
in the Animals of thofe new Countries.

Book 1. tries.   And 'tis of great confequence
in defending us from fudden accidents ;
and, efpecially when Seeing is of no ufe
to us, it fupplys its place, and gives us
feafonable warning of any imminent
danger. Befides, we fee many Animals call their fellows to them with
their Voice, which Language may
have more in it than we are aware of,
tho we don't underftand it. But if
we do but confider the vaft ufes and
neceffary occafions of Speaking on the
one fide, and Hearing on the other,
among thofe Creatures that make ufe
of their Reafon, it will fcarce feem
credible that two fuch ufeful, fuch excellent things were defigned only for
us.   For how is it poffible but that
they that are without thefe, muft be
without many other Neceffaries and
Conveniences of Life ? Or what can
they have to recompenfe this want ?
Then, if we go ftill farther, and do
but meditate upon the neat and frugal
Contrivance of Nature in making this
fame Air, by the drawing in of which
we live, by whofe Motion we fail,
and by whofe means Birds fly, for a
con-

conveyance of Sound to our ears ; and Book 1.
this Sound for the conveyance of another man's Thoughts to our Minds :
can we ever imagin that fhe has left
thofe other Worlds deftitute of fo vaft
Advantages? That they don't want *A Medium*
the means of them is certain, for their *to convey*
having Clouds in *Jupiter* puts it paft *Sound to*
doubt that they have Air too ; that *the Ear.*
being moftly formed of the Particles of
Water flying about, as the Clouds are
of them gathered into fmall Drops.
And another proof of it is, the neceffity of breathing for the prefervation of Life, a thing that feems to be
as univerfal a Dictate of Nature, as
feeding upon the Fruits of the Earth.

   As for Feeling, it feems to be given *Touch.*
upon neceffity to all Creatures that are
cover'd with a fine and fenfible Skin,
as a Caution againft coming too near
thofe things that may injure or incom-
mode them : and without it they
would be liable to continual Wounds,
Blows and Bruifes. Nature feems to
have been fo fenfible of this, that fhe
has not left the leaft place free from
fuch a perception. Therefore it's probable

Book 1. bable that the Inhabitants of thofe
Worlds are not without fo neceſſary a
Defence, and fo fit a Preſervative a-
gainſt Dangers and Miſhaps.

Smell and   And who is there that doth not fee
Taſt.   the inevitable neceſſity for all Creatures that live by feeding to have both
Taſt and Smell, that they may diſtin-
guiſh thoſe things that are good and
nouriſhing, from thoſe that are miſ-
chievous and harmful? If therefore
we allow the Planetary Creatures to
feed upon Herbs, Seeds, or Fleſh, we
muſt allow them a diſtinguiſhing Taſt
and Smell too, that they may chuſe or
refuſe any thing according as they find
it likely to be advantagious or noxious
to them.

I know that it hath been a queſtion
with many, whether there might not
have been more Senſes than thoſe five.
Their Sen- If we ſhould allow this, it might ne-
ſes not ve- vertheleſs be reaſonably doubted, whe-
ry different
from ours. ther the Senſes of the Planetary Inhabitants are much different from ours.
I muſt confeſs, I cannot deny but
there might poſſibly have been more
Senſes; but when I conſider the Uſes
of

of thofe we have, I cannot think but Book 1. they would have been fuperfluous. ～～ The Eye was made to difcern near and remote Objects, the Ear to give us notice of what our Eyes could not, either in the dark or behind our back: Then what neither the Eye nor the Ear could, the Nofe was made (which in Dogs is wonderfully nice) to warn us of. And what efcapes the notice of the other four Senfes, we have Feeling to inform us of the too near approaches of, before it can do us any mifchief. Thus has Nature fo plentifully, fo perfectly provided for the neceflary prefervation of her Creatures here, that I think fhe can give nothing more to thofe there, but what will be needlefs and fuperfluous. Yet the Senfes were not wholly defign'd for ufe: but Men from all, and all other Animals from fome of them, reap Pleafure as well as Profit, as from the Taft in delicious Meats; from the Smell in Flowers and Perfumes; from the Sight in the contemplation of beauteous Shapes and Colours; from the Hearing in the fweetnefs and harmony

mony of Sounds ; from the Feeling in Venery, unlefs you pleafe to count that for a particular Senfe by it felf. *They have* Since it is thus, I think 'tis but reafo- *Pleafure* nable to allow the Inhabitants of the *arifing* Planets thefe fame advantages that we *from the* have from them.   For upon this con- *Senfes.* fideration only, how much happier and eafier a man's Life is render'd by the enjoyment of them, we muft be obliged to grant them thefe Bleffings, except we would ingrofs every thing that is good to our felves, as if we were worthier and more deferving than any elfe. But moreover, that Pleafure which we perceive in eating or in copulation, feems to be a neceffa- ry and provident Command of Na- ture, whereby it tacitly compels us to the prefervation and continuance of our Life and Kind. It is the fame in Beafts. So that both for their hap- pinefs and prefervation it's very proba- ble the reft of the Planets are not with- out it. Certainly when I confider all thefe things, how great, noble, and ufeful they are ; when I confider what an admirable Providence it is
that

that there's fuch a thing as Pleafure in Book 1. the World, I can't but think that our Earth, the fmalleft part almoft of the Univerfe, was never defign'd to monopolize fo great a Bleffing. And thus much for thofe Pleafures which affect our bodily Senfes, but have little or no relation to our Reafon and Mind. But there are other Pleafures which Men enjoy, which their Soul only and Reafon can relifh : fome airy and brisk, others grave and folid, and yet neverthelefs Pleafures, as arifing from the Satisfaction which we feel in Knowlege and Inventions, and fearches after Truth, of which whether the Planetary Inhabitants are not partakers, we fhall have an opportunity of enquiring by and by.

There are fome other things to be confider'd firft, in which it's probable they have fome relation to us. That the Planets have thofe Elements of Earth, Air, and Water, as well as we, I have already made not unlikely. Let us now fee whether they may not have Fire too : which is not fo properly call'd an Element, as a rapid
Mo-

Book 1. Motion of the Particles in the inflam-
mable Body. But be it what it will,
All the there are many Arguments for their
Planets not being without it. For this Earth
have Fire. is not fo truly call'd the Place of Fire
as the Sun : and as by the heat of that
all Plants and Animals here thrive and
live ; fo, no doubt, is it in the other
Planets. Since then Fire is caufed by
a moft intenfe and vigorous Heat, it
follows that the Planets, efpecially
thofe nearer the Fountain of it, have
their proportionate degrees of Heat
and Fire. And when there are fo
many ways of its Produ&ion, as by
the colle&ion of the Rays of the Sun,
by the refle&ion of Mirrors, by the
ftriking of Flint and Steel, by the rub-
bing of Wood, by the clofe loading of
moift Grafs, by Lightning, by the
eruptions of Mountains and Volcanos,
it's ftrange if neither Art fhould have
produc'd it, nor Nature effe&ed it
there by one of thefe many means.
Then how ufeful and neceffary is it to
us ? By it we drive away Cold, and
fupply the want of the Sun in thofe
Countries where his oblique Rays
make

make a lefs vigorous Impreffion, and Book 1.
fo keep a great part of the Earth from
being an uninhabited Defart : which
is equally neceffary in all the Planets,
whether we allow them Succeffion of
Seafons, or a perpetual Spring and
Æquinox : for even then the Countries near the Pole would receive but
little advantage from the Heat of the
Sun. By the help of this we turn the
night into day, and thereby make a
confiderable addition to the fhortnefs
fo our Lives. Upon all thefe accounts
I muft not let this Earth of ours enjoy
it all alone, and exclude all the other
Planets from fo advantageous and fo
profitable a Gift.

But perhaps it may be asked as well
concerning Brutes as rational Creatures, and of their Plants and Trees
too, whether they are proportionably
larger or lefs than ours. For if the *The bignefs*
Magnitude of the Planets was to be the *of their Creatures*
Standard of their meafure, there would *not rightly*
be Animals in *Jupiter* ten or fifteen *guejt at ty*
times larger than Elephants, and as *the bignefs of the Pla-*
much longer than our Whales. And *nets.*
then their Men muft be mere *Goliahs*,

in

in refpect of our Pygmifhips. Now tho I don't fee any fo great abfurdity in this as to make it impoffible, yet there is no reafon to think it is really fo, feeing Nature has not always ty'd her felf to thofe Rules which we have thought more convenient for her : for example, the magnitude of the Planets is not anfwerable to their diftances from the Sun ; but *Mars*, tho more remote, is far lefs than *Venus :* and *Jupiter* turns round his Axis in ten hours, when the Earth which is much lefs than him, fpends 24. But fince Nature, perhaps fome body will fay, has not obferv'd fuch a Regularity in the proportion of things, for ought we know we may have a Race of Pygmies about the bignefs of Frogs and Mice, poffefs'd of the Planets. But I fhall fhow that this is very improbable by and by.

In the Pla-
nets are
many forts
of rational
Creatures
as well as
here.

There may arife another Queftion, whether there be in the Planets but one or more forts of rational Creatures poffefs'd of different degrees of Reafon and Senfe. There is fomething not unlike this to be obferv'd among us,

us. For to pafs by thofe who have
human Shape (altho fome of them
would very well bear that enquiry
too) if we do but confider fome forts
of Beafts, as the Dog, the Ape, the
Beaver, the Elephant, nay fome Birds
and Bees, what Senfe and Underftanding they are mafters of, we fhall be
forc'd to allow, that Man is not the
only rational Animal. For we difcover fomewhat in them of Reafon inde-
pendent on, and prior to all teaching
and practice.

But ftill no body can doubt, but
that the Underftanding and Reafon of
Man is to be prefer'd to theirs as being comprehenfive of innumerable
things, indued with an infinite memory of what's paft, and capable of providing againft what's to come. That
there is fome fuch rational Creature in
the other Planets, which is the Head
and Sovereign of the reft, is very reafonable to believe: for otherwife,
were many endued with the fame
Wifdom and Cunning, we fhould have
them always doing mifchief, always
quarrelling and fighting one another
for

for Empire and Sovereignty, a thing that we feel too much of where we have but one such Creature. But to let that pass, our next Enquiry shall be concerning those Animals in the Planets which are furnish'd with the greatest Reason, whether it's possible to know wherein they employ it, and whether they have made as great advances in Arts and Knowlege as we in our Planet. Which deserves most to be consider'd and examin'd of any thing belonging to their nature; and for the better performance of it we must take our rise somewhat higher, and nicely view the Lives and Studies of Men.

And in those things wherein Men provide and take care only of what's absolutely necessary for the preservation of their Life; in defending themselves from the Injuries of the Air; in securing themselves against the Incursions of Enemies by Walls; and against Fraud and Disturbances by Laws; in educating their Children, and providing for themselves and them: In all these I can see no great reason

reaſon that Man has to boaſt of the Book 1.
preeminency of his Reaſon above
Beaſts and other Animals. For moſt
of theſe things they perform with
greater eaſe and art than us, and
ſome of them they have no need of.
For that ſenſe of Virtue and Juſtice in
which Man excels, of Friendſhip,
Gratitude and Honeſty, of what uſe
are they, but either to put a ſtop to
the Wickedneſs of Men, or to ſecure
us from mutual Aſſaults and Injuries,
a thing wherein the Beaſts want no
Guide but Nature and Inclination?
Then if we ſet before our eyes the
manifold Cares, the diſturbances of
Mind, the reſtleſs Deſires, the dread
of Death, that are the reſult of this
our Reaſon ; and compare them with
that eaſy, quiet, and harmleſs Life
which other Animals enjoy, we ſhould
be apt to wiſh a change, and conclude
that they, eſpecially Birds, liv'd with
more pleaſure and happineſs than
Man could with all his Wiſdom. For
they have as great a guſto of bodily
Pleaſures as we, let the new Philoſo-
phers ſay what they will, who would
have

have them go for nothing but Clocks and Engines of Flefh; a thing which Beafts fo plainly confute by crying and running away from a ftick, and all other aƈtions, that I wonder how any one could fubfcribe to fo abfurd and cruel an Opinion. Nay I can fcarce doubt but that Birds feel no fmall pleafure in their eafy, fmooth failing through the Air; and would much more if they but knew the advantages it hath above our flow and *Men chiefly* laborious Progreffion. What is it *differ from* then after all that fets human Reafon *Beafts in* above all other, and makes us prefera-*the ftudy of* ble to the reft of the Animal World? *Nature.* Nothing in my mind fo much as the contemplation of the Works of God, and the ftudy of Nature, and the improving thofe Sciences which may bring us to fome knowlege in their Beauty and Variety. For without Knowlege what would be Contemplation? And what difference is there between a Man, who with a carelefs fupine negligence views the Beauty and Ufe of the Sun, and the fine golden Furniture of the Heaven, and one who

who with a learned Nicenefs fearches
into their Courfes; who underftands
wherein the Fixt Stars, as they are
call'd, differ from the Planets, and
what is the reafon of the regular Vi-
ciffitude of the Seafons; who by found
reafoning can meafure the magnitude
and diftance of the Sun and Planets?
Or between fuch a one as admires perhaps the nimble Activity and ftrange
Motions of fome Animals, and one
that knows their whole Structure, underftands the whole Fabrick and Ar-
chitecture of their Compofition? If
therefore the Principle we before laid
down be true, that the other Planets
are not inferior in dignity to ours,
what follows but that they have Crea- *They have*
tures not to ftare and wonder at the *Aftronomy.*
Works of Nature only, but who em-
ploy their Reafon in the examination
and knowlege of them, and have made
as great advances therein as we have?
They do not only view the Stars, but
they improve the Science of Aftronomy : nor is there any thing can make
us think this improbable, but that fond
conceitednefs of every thing that we
call

Book 1. call our own, and that pride that is too
natural to us to be eafily laid down.
But I know fome will fay, we are a
little too bold in thefe Affertions of
the Planets, and that we mounted hither by many Probabilities, one of
which, if it chance to be falfe, and
contrary to our fuppofition, would,
like a bad Foundation, ruin the whole
Building, and make it fall to the
ground. But I would have them to
know, that all I have faid of their
Knowlege in Aftronomy, has proofs
enough, antecedent to thofe we now
produc'd. For fuppofing the Earth,
as we did, one of the Planets of equal
dignity and honor with the reft,
who would venture to fay, that no
where elfe were to be found any that
enjoy'd the glorious fight of Nature's
Opera? Or if there were any fellow-
Spectators, yet we were the only ones
that had dived deep into the fecrets
and knowlege of it? So then here's a
proof not fo far fetch'd for the Aftronomy of the Planets, the fame which
we ufed for their having rational Creatures, and enjoying the other advantages

tages we before talk'd of, which ferves
at the fame time for the confirmation
of our former Conjectures. But if
Amazement and Fear at the Eclipfes
of the Moon and Sun gave the firft
occafion to the ftudy of Aftronomy,
as they fay it did, then it's almoft impoffible that *Jupiter* and *Saturn* fhould
be without it; the Argument being of
much greater force in them, by reafon of the daily Eclipfes of their
Moons, and the frequent ones of the
Sun to their Inhabitants. So that if a
Perfon difinterefted in his Judgment,
and equally ignorant of the Affairs of
all the Planets, were to give his Opinion in the matter, I don't doubt he
would give the caufe for Aftronomy to
thofe two Planets rather than us.

This fuppofition of their Knowlege
and Ufe of Aftronomy in the Planetary Worlds, will afford us many new
Conjectures about their manner of life,
and their ftate as to other things.

For, Firft: No Obfervations of *And all its*
the Stars that are neceffary to the *fubfervient*
knowlege of their Motions, can be *Arts.*
made without Inftruments; nor can
thefe

Book.i. thefe be made without Metal, Wood, or fome fuch folid Body. Here's a neceffity of allowing them the Carpenters Tools, the Saw, the Ax, the Plane, the Mallet, the File : and the making of thefe requires the ufe of Iron, or fome equally hard Metal. *Geometry and Arithmetick :* Again, thefe Inftruments can't be without a Circle divided into equal Parts, or a ftreight Line into unequal. Here's a neceffity for introducing Geometry and Arithmetick. Then the *And Writing.* neceffity in fuch Obfervations of marking down the Epochas or Accounts of Time, and of tranfmitting them to Pofterity, will force us to grant them the Art of Writing; I won't fay the fame with ours which is commonly ufed, but I dare affirm not more ingenious or eafy. For how much more ready and expeditious is our way, than by that multitude of Charaĉters ufed in *China* ; and how vaftly preferable to Knots tied in Cords, or the Piĉtures in ufe among the barbarous People of *Mexico* and *Peru* ? There's no Nation in the World but has fome way or other of writing and marking down their

their Thoughts : So that it's no wonder if the Planetarians have been taught
it by that great School-miſtreſs Neceſſity, and apply it to the ſtudy of
Aſtronomy and other Sciences. In
Aſtronomical matters the neceſſity of
it is moreover apparent from hence,
that the motion of the Stars is as
'twere to be fancied and gueſs'd at in
different Syſtems, and theſe Syſtems
to be continually improved and cor-
rected, as later and more exact Obſervations ſhall convince the old ones of
faults : all which can never be deli-
ver'd down to ſucceeding Generations,
unleſs we make uſe of Letters and Figures.

But for all our large and liberal al-
lowances to theſe Gentlemen, they
will ſtill be behind-hand with us. For
we have ſo certain a knowlege of the *And Op-*
true Syſtem and Frame of the Uni-*ticks.*
verſe ; we have ſo admirable an Invention of Teleſcopes to help our failing
Eye-ſight in the view of the bigneſs
and different forms of the Planetary
Bodies, in the diſcovery of the Mountains, and the Shadows of them on
the

Book I. the Surface of the Moon, in the
bringing to light an innumerable multitude of Stars otherwife invifible, that
we muft neceffarily be far their Mafters in that Knowlege.   What muft
I do here? I could find in my heart
(and I can fee no reafon why I may
not, except it be to flatter and com-
plement our felves in being the only
People that have the advantage of fuch
excellent Inventions) either to allow
thefe Planetary Inhabitants fuch fharp
Eyes as not to need them, or elfe the
ufe of Glaffes to help the deficiency of
their Sight.   And yet I dare not, for
fear People fhould be fo difturbed at
the ridiculous Extravagancy of fuch an
Opinion, as to take the meafure of
my other Conjectures by it, and hifs
them all off, upon the account of this
alone.

Thefe Sci-   But fome body may perhaps object,
ences not and that not without reafon at firft
contrary to fight, that the Planetarians it's likely
Nature.   are deftitute of all refined Knowlege,
juft as the Americans were before they
had Commerce with the Europeans.
For if one confiders the Ignorance of
those

those Nations, and of others in *Asia* and *Africa* equally barbarous, it will appear as if the main design of the Creator in placing Men upon the Earth was that they might live, and, in a juft fenfe of all the Bleffings and Pleafure they enjoy, worfhip the Fountain of their Happinefs; but that fome bold fellows have leapt over the bounds of Nature, and made fearches into thofe forbidden depths only out of an affectation of knowing more than they were made for. There does not want an Anfwer for thefe Men. For God could not but forefee the advances Men would make, in their enquiring into the Affairs of Heaven: that they would difcover Arts ufeful and advantageous to Life: that they would crofs the Seas, and dig up the Bowels of the Earth. Nothing of all this could happen contrary to the Mind and Knowlege of the Infinite Author of all things. And if he forefaw thefe things would be, he fo appointed and deftin'd them to human kind. And the Studies of Arts and Sciences cannot be faid to be contrary

trary to Nature, since in the search thereof they are employ'd: especially if we consider the natural desire and love of Knowlege, rooted in all men. For it's impossible this should have been given them upon no design or account. But they will urge, that if such a Knowlege is natural, if we were born for it, why are there so very few, especially in Astronomy, that prosecute these Studies? For *Europe* is the only Quarter of the Earth in which there have been any advancements made in Astronomy. And as for the Judicial Astrology, that pretends to foretel what is to come, it is such a ridiculous, and oftentimes mischievous Folly, that I do not think it fit to be so much as named. And even in *Europe,* not one in a hundred thousand meddles with these Studies. Besides, its Original and Rise is so late, that many Ages were past before the very first Rudiments of Astronomy or Geometry (which is necessary to the learning of it) were known. For every body is acquainted almost with its first beginnings in *Egypt* and *Greece.* Add to this,

this, that 'tis not yet above fourfcore years fince the bungling Epicycles were difcarded, and the true and eafy plain Motion of the Planets was difcover'd. For the fatisfaction of thefe Scruples, to what we faid before, concerning the Fore-knowlege of God, may be added this; That God never defign'd we fhould come into the World Aftronomers or Philofophers; thefe Arts are not infus'd into us at our birth, but were order'd, in long tracts of Time, by degrees to be the rewards and refult of laborious Diligence: efpecially thofe Sciences which are now in debate, are fo much the more difficult and abftrufe, that their late Invention and flow Progrefs are fo far from being a wonder, that it is rather ftrange they were ever difcover'd at all. There are but few, I acknowlege one or two perhaps, in an age, that purfue them, or think them their bufinefs: but their number will be very confiderable if we take in thofe that have liv'd in all the ages in which Aftronomy hath flourifhed: and no body can deny them that happinefs and contentment which
they

they have pretended to above all others. In fine, it was fufficient that fo fmall a number fhould make it their ftudy, fo that the Profit and Advantage of their Inventions might but fpread it felf over all the World. Since then the Inhabitants of this Earth, let them be never fo few, have had Parts and Genius fufficient for the attainment of this Knowlege; and there's no reafon to think the Planetarians lefs ingenious or happy than our felves; we have gained our point, and 'tis probable that they are as fkilful Aftronomers as we can pretend to be. So that now we may venture to deduce fome Confequences from fuch a Suppofition.

We have before fhow'd the neceffary Dependence and Connexion, not only of Geometry and Arithmetick, but of mechanical Arts and Inftruments with this Science. This leads us naturally to the enquiry how they can ufe thefe Inftruments and Engines for the obfervation of the Stars, how they can write down fuch their Obfervations, and perform other things which we do with our hands. So that

that we muſt neceſſarily give them Book 1.
hands, or ſome other Member, as ᨶᨶᨶ
convenient for all thoſe uſes, inſtead of *They have*
them. I know an antient Philoſopher *Hands.*
laid ſuch ſtreſs upon the uſe and conveniency of the hands, that he made
no ſcruple to affirm, they were the
cauſe and foundation of all our Knowlege. By which, I ſuppoſe, he meant
no more, than that without their help
and aſſiſtance men could never arrive
to the improvement of their Minds in
natural Knowlege : And truly not
without reaſon. For ſuppoſe inſtead
of them they had had Hoofs like
Horſes or Bullocks given them, they
might have laid indeed the model and
deſign of them in their Head, but
they would never have been able to
have built Cities and Houſes. They
would have had no Subje&t of Diſ-
courſe but what belong'd to their Vic-
tuals, Marriages, or Self-preſervation.
They would have been void of all
Knowlege and Memory, and indeed
would have been but one degree diſtant from brute Beaſts. What could
we invent or imagine that could be ſo
exactly

exactly accommodated to all the design'd uses as the Hands are? Shall we
give them an Elephants Probofcis.
'Tis true, thefe Beafts can lay hold of,
or throw any thing, can take up even
the fmalleft things from the Ground,
and can perform fuch admirable feats
with it, that it has not very improper-
ly been call'd their Hand, tho indeed
it is nothing but a Nofe fomewhat longer than ordinary. Nor do Birds
fhow lefs Art and Defign in the ufe of
their Bills in the picking up their Meat,
and the wonderful compofure of their
Nefts. But all this is nothing to
thofe Conveniences the Hand is fo
admirably futed to; nothing to that
amazing contrivance in its capacity
of being ftretch'd, or contracted, or
turned to any part as occafion fhall require. And then, to pafs by that nice
Senfe that the ends of the Fingers are
endued with, even to the feeling and
diftinguifhing moft forts of Bodies in
the dark, what Wifdom and Art is
fhow'd in the difpofition of the Thumb
and Fingers, fo as to take up or keep
faft hold of any thing we pleafe? Either

ther then the Gentlemen that live there Book 1.
muft have Hands, or fomewhat equal-
ly convenient, which is no eafy matter; or elfe we muft fay that Nature
has been kinder not only to us, but
even to Squirrels and Monkeys than
them.

That they have Feet fcarce any one *And Feet.*
can doubt, that does but confider
what we faid but juft now of the different methods of Progreffion, which
it's hard to imagin can be perform'd
any other ways than what we there
recounted. And, of all thofe, there's
none can agree fo well with the ftate
of the Planetarians, as that that we
here make ufe of. Except (what is
not very probable, if they live in Society, as I fhall fhow they do) they
have found out the art of flying in
fome of thofe Worlds.

The Stature and Shape of Men here *That they*
does fhow forth the Divine Provi- *are upright*
dence fo much in its being fo fitly
adapted to its defign'd Ufes, that it is
not without reafon that all the Philofophers have taken notice of it nor
without probability that the Planetarians

rians have their Eyes and Countenance upright, like us, for the more convenient and eafy Contemplation and Obfervations of the Stars. And the Wifdom of the Creator is fo obfervable, fo praifeworthy in the pofition of the other Members ; in the convenient fituation of the Eyes, as Watches in the higher Region of the Body ; in the removing of the more uncomly parts out of fight as 'twere ; that we cannot but think he has almoft obferved the fame Method in the Bodies of thofe remote Inhabitants. Nor *It follows* does it follow from hence that they *not there-* muft be of the fame fhape with us. *fore that* For there is fuch an infinite poffible *they have* variety of Figures to be imagined, that *the fame* *fhape with* both the Oeconomy of their whole Bo*us.* dies, and every part of them, may be quite diftinct and different from ours. How warmly and conveniently are fome Creatures clothed with Wool, and how finely are others deck'd and adorn'd with Feathers? Perhaps among the rational Creatures in the Planets there may fome fuch diftinction be obferv'd in their Garb and Covering ;

vering; a thing in which Men are apt to envy the happiness of Beasts, tho perhaps without reason. For men might be born naked, only perhaps for the employment and exercising their Wits, in the inventing and making that Attire that Nature had made necessary for them. And 'tis this necessity that has been the greatest, if not only occasion of all the Trade and Commerce of all the Mechanical Inventions and Discoveries that we are masters of. Besides, Nature might have another great Conveniency in her eye, by bringing men into the World naked, namely, that they might accommodate themselves to all places of the World, and go thicker or thinner cloth'd, according as the Season and Climate they liv'd in required. There may still be a greater difference between us and them; for there is a sort of Animals in the World, as Oysters, Lobsters, and Crab-fish, whose Flesh is on the inside of their Bones as 'twere. What if the Planetarians should be such? O no, some body will say, it would be a hideous sight,

so

Book 1. fo ugly, that Nature has not made any but her refuse and meaner Creatures of fuch an odd Compofition. As for that, I fhould not be at all moved with their ugly fhape, if it were not, that hereby they would be deprived of that quick eafy motion of their Hands and Fingers, which is fo ufeful and necef- fary to them.

A rational Soul may inhabit a- nother Shape than ours. For 'tis a very ridiculous opinion, that the common people have got a- mong them, that it is impoffible a ra- tional Soul fhould dwell in any other fhape than ours. And yet as filly as 'tis, it has been the occafion of many Philofophers allowing the Gods no other fhape ; nay, the Foundation of a Sect among the Chriftians, that from hence have the name of *Anthropo- morphites*. This can proceed from nothing but the Weaknefs, Igno- rance, and Prejudice of Men ; as well as that too of humane Figure being the handfomeft and moft excellent of all others, when indeed it's nothing but a being accuftomed to that figure that makes us think fo, and a conceit that we and all other Animals natu- rally

rally have, that no fhape or colour can Book 1.
be fo good as our own. Yet methinks ∿
this fancy has fuch a rule upon my
mind, that 1 cannot without horror
and impatience fuffer any other figure
for the habitation of a reafonable Soul.
For when I do but reprefent to my I-
magination or Eyes a Creature like a
Man in every thing elfe, but that has
a Neck four times as long, and great
round fawcer Eyes five or fix times as
big, and farther diftant, I cannot look
upon't without the utmoft averfion,
altho at the fame time I can give no
account of my Diflike.

As I was talking fomewhat above *The Pla-*
of the Stature of the Planetary Inha- *netarians*
bitants, I hinted that 'twas improba- *not lefs*
ble they fhould be lefs than we are. *than we.*
For it's likely, that as our Bodies are
made in fuch a proportion to our
Earth, as to render us capable of tra-
velling about it, and making Obfervations upon its bulk and figure, the
fame Order is obferv'd in the Inhabitants of the other Planets, except here
too our Pride put in for our Preeminence. Then feeing we have before
allow'd

Book 1. allow'd them Aftronomy and Obfervations, we muft give them Bodies
and Strength fufficient for the ruling
their Inftruments, and the erecting
their Tubes and Engines. And for
this the larger they are the better. For
if we fhould make them little Fellows
about the bignefs of Rats or Mice,
they could neither make fuch Obfervations as are requifite ; nor fuch Inftruments as are neceffary to thofe Obfervations. Therefore we muft fuppofe
them larger than, or at leaft equal to
our felves, efpecially in *Jupiter* and
Saturn, which are fo vaftly bigger
than the Planet which we inhabit.

They live in Society. Aftronomy, we faid before, could
never fubfift without the writing
down the Obfervations : nor could
the Art of Writing (any more than
the Carpenters and Founders) ever be
found out except in a Society of reafonable Creatures, where the neceffi-
ties of Life forc'd them upon Invention : So that what I promis'd to prove
follows from hence, namely, that the
Planetarians muft in this be like us,
that they maintain a Society and Fel-
lowfhip

lowſhip with, and afford mutual Aſ-
ſiſtances and Helps to one another.
Hereupon we muſt allow them a ſet-
tled, not a wandring *Scythian* way of
living, as more convenient for men in
ſuch circumſtances. But what then?
Shall they have every thing elſe proper
for ſuch a manner of living granted
them too? Shall they have their Go-
vernours, Houſes, Cities, Trade, and
Bartering? Why not? when even the
barbarous People of *America* and other
places were at their firſt diſcovery
found to have ſomewhat of that nature in uſe among them. I won't ſay,
that things muſt be the ſame there as
they are here. We have many that
may very well be ſpared among rational Creatures, and were deſign'd only
for the preſervation of Society from all
Injury, and for the curbing of thoſe
men who make an ill uſe of their Reaſon to the detriment of others. Perhaps in the Planets they have ſuch plen-
ty and affluence of all good things, as
they neither need or deſire to ſteal from
one another ; perhaps they may be ſo
juſt and good as to be at perpetual
Peace,

Book 1. Peace, and never to lie in wait for, or
take away the Life of their Neigh-
bour: perhaps they may not know
what Anger or Hatred are ; which we
to our coft and mifery know too too
well. But ftill it's more likely they
have fuch a medly as we, fuch a mixture of good with bad, of wife with
fools, of war with peace, and want not
that Schoolmiftrefs of Arts Poverty.
For thefe things are of no fmall ufe :
and if there were no other, 'twould
be reafon enough that we are as good
Men as themfelves.

They enjoy   What I am now going to fay may
the plea-   feem fomewhat more bold, and yet is
fures of   not lefs likely than the former. For
Society.   if thefe new Nations live in Society,
as I have pretty well fhow'd they do,
'tis fomewhat more than probable that
they enjoy not only the Profit, but the
Pleafures arifing from fuch a Society :
fuch as Converfation, Amours, Jeft-
ing, and Sights. Otherwife we fhould
make them live like fo many *Catos*,
without Diverfion or Merriment ; we
fhould deprive them of the great
Sweetnefs of Life, which it can't well
be

be without, and give our felves fuch Book 1.
an advantage over them as Reafon
will by no means admit of.

But to proceed to a farther Enquiry
into their Bufinefs and Employment,
let's confider what we have not already mention'd, wherein they may bear
any likenefs to us. And firft we have
good reafon to believe they build themfelves Houfes, becaufe we are fure
they be not without their Showers.
For in *Jupiter* have been obferv'd
Clouds, big no doubt with Vapors
and Water, which hath been proved
by many other Arguments, not to be
wanting in that Planet. They have
then their Rain, for otherwife how
could all the Vapors drawn up by the
heat of the Sun be difpofed of? and
their Winds, for they are caufed only
by Vapors diffolved by heat, and it's
plain that they blow in *Jupiter* by the *They have*
continual motion and variety of the *Houfes to*
Clouds about him. To protect them- *fecure 'em*
felves from thefe, and that they may *ther.*
pafs their Nights in quiet and fafety,
they muft build themfelves Tents or
Huts, or live in holes of the Earth.
For

Book 1. For I dare not affront the Pride of Men so much as to say, they are as good Architects, have as noble Houses, and as stately Palaces as our selves. And good now who are we? Why a company of mean fellows living in a little corner of the World, upon a Ball ten thousand times less than *Jupiter* or *Saturn.* And yet we forsooth must be the only skilful People at Building : and all others must be our Inferiours in the knowlege of uniform Symmetry ; and not be able to raise Towers and Pyramids as high, magnificent, and beautiful, as our selves. For my part, I see no reason why they may not be as great Masters at it as we are, and have the use of all those *Arts* subservient to it, as Stone-cutting and Brick-making, and whatsoever else is necessary for it, as Iron, Lead and Glass ; or ornamental to it, as Gilding and Picture.

If their Globe is divided like ours, between Sea and Land, as it's evident it is (else whence could all those Vapors in *Jupiter* proceed?) we have great reason to allow them the Art of Navi-

Navigation, and not proudly ingrofs
fo great, fo ufeful a thing to our felves.
Efpecially confidering the great advantages *Jupiter* and *Saturn* have for failing, in having fo many Moons to di-
rect their Courfe, by whofe guidance
they may attain eafily to the Knowlege that we are not Mafters of, of the
Longitude of Places. And what a
troop of other things follow from this
allowance? If they have Ships, they
muft have Sails and Anchors, Ropes,
Pullies, and Rudders, which are of
particular ufe in directing a Ship's
Courfe againft the Wind, and in failing different ways with the fame Gale.
And perhaps they may not be without
the ufe of the Compafs too, for the
magnetical matter, which continually
paffes through the Pores of our Earth,
is of fuch a nature, that it's very pro- *They have*
bable the Planets have fomething like *Navigati-*
it. But there's no doubt but that they *on, and all*
Arts fub-
muft have the Mechanical Arts and *fervient.*
Aftronomy, without which Naviga-
tion can no more fubfift, than they
can without Geometry.

But

Book 1.   But Geometry ſtands in no need of
being proved after this manner.   Nor
doth it want aſſiſtance from other Arts
which depend upon it, but we may
have a nearer and ſhorter aſſurance of
their not being without it in thoſe
Earths. For that Science is of ſuch ſin-
gular worth and dignity, ſo peculiarly
imploys the Underſtanding, and gives
it ſuch a full comprehenſion and infalli-
As Geome- ble certainty of Truth, as no other
try. Knowlege can pretend to : it is moreover of ſuch a nature, that its Principles and Foundations muſt be ſo immutably the ſame in all times and
places, that we cannot without In-
juſtice pretend to monopolize it,
and rob the reſt of the Univerſe of
ſuch an incomparable Study. Nay
Nature it ſelf invites us to be Geome-
tricians : it preſents us with Geo-
metrical Figures, with Circles and
Squares, with Triangles, Polygones,
and Spheres, and propoſes them as
it were to our conſideration and ſtudy,
which abſtracting from its Uſeful-
neſs, is moſt delightful and raviſhing.
Who can read *Euclid*, or *Apollonius*,
about

about the Circle, without admiration?
or *Archimedes* of the Surface of the
Sphere, and Quadrature of the Para-
bola without amazement? or confider
the late ingenious Difcoveries of the
Moderns with Boldnefs and Uncon-
cernednefs? And all thefe Truths are
as naked and open, and depend upon
the fame plain Principles and Axioms
in *Jupiter* and *Saturn* as here, which
makes it not improbable that there are
in the Planets fome who partake with
us in thefe delightful and pleafant Studies. But what's the greateft Argument with me, that there are fuch is
their ufe, I had almoft faid neceffity,
in moft Affairs of humane Life. Now
we are got thus far, what if we fhould
venture fomewhat farther, and tell
you, that they have our Inventions of
the Tables of Sines, of Logarithms,
and Algebra: I know I fhould be
laugh'd at for an idle Difcoverer of nothing but ridiculous Whimfies, and yet
there's no reafon but the old one, of
our being better than all the World,
to hinder them from being as happy
in their Difcoveries, and as ingenious
in

in their Inventions as we our felves are.

They have Mufick.   It's the fame with Mufick as with Geometry, it's every where immutably the fame, and always will be fo. For all Harmony confifts in Concord, and Concord is all the World over fixt according to the fame invariable meafure and proportion. So that in all Nations the difference and diftance of Notes is the fame, whether they be in a continued gradual progreffion, or the voice makes skips over one to the next. Nay very credible Authors report, that there's a fort of Bird in *America*, that can plainly fing in order fix mufical Notes : whence it follows that the Laws of Mufick are unchangeably fix'd by Nature, and therefore the fame Reafon holds valid for their Mufick, as we e'en now propofed for their Geometry. For why, fuppofing other Nations and Creatures, endued with Reafon and Senfe as well as we, fhould not they reap the Pleafures arifing from thefe Senfes as well as we too ? I don't know what effect this Argument, from the immutable nature of thefe Arts,

Arts, may have upon the Minds of Book 1. others ; I think it no inconfiderable or contemptible one, but of as great Strength as that which I made ufe of above to prove that the Planetarians had the fenfe of Seeing.

But if they take delight in Harmony, 'tis twenty to one but that they have invented mufical Inftruments. For, if nothing elfe, they could fcarce help lighting upon fome or other by chance ; the found of a tight String, the noife of the Winds, or the whiftling of Reeds, might have given them the hint. From thefe fmall beginnings they perhaps, as well as we, have advanced by degrees to the ufe of the Lute, Harp, Flute, and many ftring'd Inftruments. But altho the Tones are certain and determinate, yet we find among different Nations a quite different manner and rule for Singing ; as formerly among the Dorians, Phrygians, and Lydians, and in our time among the French, Italians, and Perfians. In like manner it may fo happen, that the Mufick of the Inhabitants of the Planets may widely differ from

Book 1. from all thefe, and yet be very good.
But why we fhould look upon their
Mufick to be worfe than ours, there's
no reafon can be given; neither can
we well prefume that they want the
ufe of half-notes and quarter-notes,
feeing the invention of half-notes is fo
obvious, and the ufe of 'em fo agreeable to nature. Nay, to go a ftep farther, what if they fhould excel us in the
Theory and practick part of Mufick,
and outdo us in Conforts of vocal and
inftrumental Mufick, fo artificially
compos'd, that they fhew their Skill by
the mixtures of Difcords and Concords?
and of this laft fort 'tis very likely the
5*th* and 3*d* are in ufe with them.

This is a very bold Affertion, but it
may be true for ought we know, and
the Inhabitants of the Planets may pof-
fibly have a greater infight into the
Theory of Mufick than has yet bin difcover'd amongft us. For if you ask any
of our Muficians, why two or more perfect fifths cannot be us'd regularly in
compofition; fome fay 'tis to avoid
that Sweetnefs and Lufhioufnefs which
arifes from the repetition of this plea-
fing

fing Chord : Others fay, this muſt be
avoided for the fake of that variety of
Chords that are requiſite to make a
good compoſition ; and theſe Reaſons
are brought by *Cartes* and others. But
an Inhabitant of *Jupiter* or *Venus* will
perhaps give you a better reaſon for this,
viz. becauſe when you paſs from one
perfect fifth to another, there is ſuch a
change made as immediately alters
your Key, you are got into a new
Key before the Ear is prepared for it,
and the more perfect Chords you uſe
of the ſame kind in Confecution, by ſo
much the more you offend the Ear by
theſe abrupt Changes.

Again, one of theſe Inhabitants will
tell you how it comes about, that in a
Song of one or more Parts, the Key
cannot be kept ſo well in the ſame a-
greeable Tenor, unleſs the intermediate Cloſes and Intervals be ſo temper'd,
as to vary from their uſual Proportions, and thereby to bear a little this
way or that, in order to regulate the
Scale. And why this Temperature is
beſt in the Syſtem of the Strings, when
out of the fifth the fourth part of a
Comma

Book 1. Comma is ufually cut off; This fame thing I have formerly fhew'd at large.

But for the regulating the Tone of the Voice (as I before hinted) that may admit of a more eafy proof, and we fhall give you an Effay of it, being unwilling ftill to put you off with my own whims: I fay therefore, if any Perfons ftrike thofe Sounds which the Muficians diftinguifh by thefe Letters, C, F, D, G, C, by thefe agreeable Intervals, altogether perfect, inter-changable, afcending and defcending with the Voice: Now this latter found C will be one Comma, or very fmall portion lower than the firft founding of C. Becaufe of thefe perfect Intervals, which are as 4 to 3, 5 to 6, 4 to 3, 2 to 3, an account is made in fuch a proportion, as 160 to 162, that is as 80 to 81, which is what they call a Comma. So that if the fame Sound fhould be repeated nine times, the Voice would fall near the matter a greater Tone, whofe proportion is as 8 to 9. But this the fenfe of the Ears by no means endures, but remembers the firft Tone, and returns to it again.

There-

Therefore we are compell'd to ufe an Book 1.
occult Temperament, and to fing thefe ⌇⌇
imperfeſt Intervals, from doing which
lefs offence arifes. And for the moft
part, all Singing wants this Temperament, as may be collected by the afore-
faid Computations. And thefe things
we have offer'd to thofe that have
fome Knowlege in Geometry.

We have fpoke of thefe Arts and
Inventions, which it is very probable
the Inhabitants of the Planets partake
of in common with us, befides which
it feems requifite to take in many other
things that ferve either for the ufe or
pleafure of their Lives. But what
thefe things are we fhall the better account for, by laying before us many of
thofe things which are found amongft
us. I have before mention'd the variety of Animals and Vegetables, which
very much differ from each other,
among which there are fome that differ but little ; and I have faid, that
there are no lefs differences in thefe
things in the Planetary Worlds.

I fhall now take a fhort view of the
Benefits we receive both from thofe
Herbs

Book 1. Herbs and Animals, and fee whether
we may not with very good reafon
conclude that the Planetarians reap as
great and as many from thofe that
their Countries afford them.

And here it may be worth our while
to take a review of the variety and
multitude of our Riches. For Trees
and Herbs do not only ferve us for
Food, they in their delicious Fruits,
thefe in their Seeds, Leaves and Roots;
but Herbs moreover furnifh us with
Phyfick, and Trees with Timber for
our Houfes and Ships. Flax, by the
means of thofe two ufeful Arts of
Spinning and Weaving, affords us
Clothing. Of Hemp or Matweed
we twift our felves Thread and fmall
Ropes, the former of which we employ in Sails and Nets, the latter in
making larger Ropes for Mafts and
Anchors. With the fweet Smells and
beauteous Colours of Flowers we feaft
our Senfes: and even thofe of them

The Ad- that offend our Noftrils, or are mif-
vantages chievous to our Bodies, are feldom
we reap
from Herbs without excellent ufes: or were made
and Ani- perhaps by Nature as a foil to fet off,
mals.

and

and make us the more value the good Book 1.
by comparing them with thefe. What ∿
vaft advantages and profit do we reap
from the Animals? The Sheep give us
Clothing, and the Cows afford us
Milk: and both of them their Flefh
for our Suftenance. Affes, Camels,
and Horfes do, what if we wanted
them we muft do ours felves, carry
our Burdens; and the laft of them we
make ufe of, either themfelves to carry us, or in our Coaches to draw us.
In which we have fo excellent, fo ufeful an Invention of Wheels, that I
can't let the Planets enjoy Society and
all its confequences, and be without
them. Whether they are Pythagore-
ans there, or feed upon Flefh as we
do, I dare not affirm any thing. Tho
it feems to be allow'd Men to feed upon whatfoever may afford them Nourifhment, either on Land, or in Water, upon Herbs, and Pomes, Milk,
Eggs, Honey, Fifh, and no lefs upon
the Flefh of many Birds and Beafts.
A ftrange thing! that a rational Creature fhould live upon the Ruin and
Deftruction of fuch a number of other
his

Book 1. his Fellow-Creatures! And yet not at all unnatural fhould it feem, fince not only he, but even Lions, Wolves, and other ravenous Beafts, prey upon Flocks of other harmlefs things, and make mere Fodder of them; as Eagles do of Pidgeons and Hares; and large Fifh of the helplefs little ones. We have different forts of Dogs for Hunting, and what our own Legs cannot, that their Nofe and Legs can help us to. But the Ufe and Profit of Herbs and Animals are not the only things they are good for, but they raife our delight and admiration when we confider their various Forms and Natures, and enquire into all their different ways of Generation: things fo infinitely multifarious, and fo delightfully amazing, that the Books of Natural Philofophers are defervedly fill'd with their Encomiums. For even in the very Infects, who can but admire the fix-corner'd Cells of the Bees, or the artificial Web of a Spider, or the fine Bag of a Silk-worm, which laft affords us, with the help of incredible Induftry, even Shiploads of foft delicate Clothing.

Clothing. This is a fhort Summary
of thofe many profitable Advantages
the animal and herbal World ferve us
with.

But this is not all. The Bowels of
the Earth too muft contribute to Man's
Happinefs. For what art and cunning does he employ in finding, in
digging, in trying Metals, and in
melting, refining, and tempering them?
What Skill and Nicety in beating,
drawing or diffolving Gold, fo as with *And from*
inconfiderable changes to make every *Metals.*
thing he pleafes put on that noble
Luftre? Of how many and admirable
ufes is Iron ? and how ignorant in all
Mechanical Knowlege were thofe Nations that were not acquainted with it,
fo as to be fain to ufe no Arms but
Bows, Clubs, and Spears, made of
Wood. Poor Weapons! There's one
thing indeed we have, which it's a
queftion whether it has done more
harm or good, and that's a devilifh
Powder made of Nitre and Brimftone.
At firft indeed it feem'd as if we had
got a more fecure Defence than former
Ages againft all Affaults, and could
eafily

Book 1. eafily guard our Towns, by the wonderful ftrength of that Invention, a-
gainft all hoftile Invafions: but now
we find it has rather encouraged them,
and at the fame time bin no fmall occafion of the decay of Valor, by ren-
dring it and Strength almoft ufelefs in
War. Had the Grecian Emperor who
faid, *Virtue was ruin'd* only when
Slings and Rams firft came into ufe,
liv'd in our days, he might well have
complain'd; efpecially of Bombs, a-
gainft which neither Art nor Nature
is of fufficient proof: but which be it
never fo ftrong, lays every thing,
Caftles and Towers, even with the
Ground. If for nothing elfe, yet upon this one account, I think we had
better have bin without the Difcovery. Yet, when we were talking of
our Difcoveries, it was not to be
pafs'd over, for the Planets too may
have their mifchievous as well as ufeful Inventions.

We are happier in the ufes for
which the Air and Water ferve us;
both of which help us in our Navigation, and furnifh us with a Strength
fuffi-

fufficient, without any labor of our Book 1.
own, to turn round our Mills and Engines; things which are of ufe to us in
fo many different Employments. For
with them we grind our Corn, and
fqueeze out our Oyl ; with them we
cut Wood, and mill Cloth, and with
them we beat our ftuff for Paper. An
incomparable Invention ! Where the
naftieft ufelefs fcraps of Linen are
made to produce fine white Sheets.
To thefe we may add the late difcovery of Printing, which not only pre-
ferves from Death Arts and Knowlege, but makes them much eafier to
be attained than before. Nor muft
we forget the Arts of Engraving and
Painting, which from mean beginnings have improv'd to that Excel-
lence, that nothing that ever fprung
from the Wit of Man can claim Preeminence to them. Nor is the way
of melting and blowing Glaffes, and
of polifhing and fpreading Quickfilver
over Mirrors, unworthy of being
mention'd, nor above all the admirable ufes that Glaffes have bin put to in
natural Knowlege, fince the invention
of

Book 1. of the Telefcope and Microfcope. And
no lefs nice and fine is the Art of making Clocks, fome of which are fo
fmall as to be no weight to the Bearer;
and others fo exact as to meafure out
The Au- the Time in as fmall Portions as any
thor in- one can defire: the improvement of
vented the both which the World owes to my
Pendulum Inventions.
for Clocks. * Inventions.

From the I might add much here of the late
difcoveries Difcoveries, moft of them of this age,
of our Age. which have bin made in all forts of
Natural Knowlege as well as in Geo-
metry and Aftronomy, as of the
weight and fpring of the Air, of the
Chymical Experiments that have
brought to light a way of making Li-
quors that fhall fhine in the dark, and
with gentle moving fhall burn of themfelves. I could tell you of the Circu-
lation of the Blood through the Veins
and Arteries, which was underftood
indeed before; but now, by the help
of the Microfcope, has an ocular De-
monftration in the Tails of fome
Fifhes: of the Generation of Animals,
which now is found to be perform'd
no otherwife than by the Seed of one
of

of the fame kind; and that in the Book 1.
Seed of the Male are difcover'd, by
the help of Glaffes, Millions of fpright-
ly little Animals, which it's probable
are the very Offspring of the Animals
themfelves : a wonderful thing, and
never before now known!

Thus have I heap'd together all *The Pla-*
nets have,
thefe late Difcoveries of our Earth : *tho not*
and now, tho perhaps fome of them *thefe fame,*
may be common to the Planetarians *yet as ufe-*
ful Inven-
with us, yet that they fhould have all *tions.*
of them is not credible. But then they
have fomewhat to make up that de-
fect, others as good and as ufeful, and
as wonderful, that we want. We have
allow'd that they may have rational
Creatures among them, and Geome-
tricians, and Muficians : we have
prov'd that they live in Societies, have
Hands and Feet, are guarded with
Houfes and Walls : yet if a Man was
but carried thither by fome powerful
Genius, fome *Pegafus,* I don't doubt
'twould be a very pretty fight, pretty
beyond all imagination, to fee the odd
ways, and the unufual manner of their
fetting about any thing, and their
ftrange

Book 1. ſtrange methods of living. But ſince there's no hopes of a *Mercury* to carry us ſuch a Journey, we ſhall e'en be contented with what's in our power : we ſhall ſuppoſe our ſelves there, and inquire as far as we can into the Aſtronomy of each Planet, and ſee in what manner the Heavens preſent themſelves to their Inhabitants. We ſhall make ſome Obſervations of the Eminence of each of them, in reſpect of their Magnitude, and number of Moons they have to wait on them ; and ſhall propoſe a new Method of coming to ſome knowlege of the incredible diſtance of the fix'd Stars. But firſt after this long Trouble we will give our Reader a breathing while.

New

New Conjectures concerning the Planetary Worlds.

BOOK the Second.

'TWAS a pretty many years
ago that I chanc'd to light upon *Athanafius Kircher's* Book,
call'd, *The Ecftatick Journey*, which
treats of the nature of the Stars, and
of all things that are to be found in the
Planets: I wonder'd to fee nothing
there of what I had often thought not
improbable, but quite other things,
nothing but a company of idle unrea-
fonable ftuff: which I was the more
confirm'd in, when, after the writing
of the former part, I ran over the
Book again. And methoughts mine
were very notable weighty Matters if
but compar'd with *Kircher's*. That
other People may be fatisfied in this,
and fee how vainly thofe, who caft off
the only Foundations of Probability in
fuch matters, which we have all the
way made ufe of, pretend to philofo-
phize

Book 2. phize in this cafe, I don't care if I beftow
ſome few Refleƈtions upon that Book.

Kircher's That ingenious Man ſuppoſing him-
Journey in ſelf carry'd by ſome Angel through
Ecſtacy ex- the vaft ſpaces of Heaven, and round
amin'd. the Stars, tells us, he ſaw a great many things, ſome of which he had out
of the Books of Aftronomers, the reſt
are the produƈt of his own Fancy and
Thoughts. But, before he enters upon his Journey, he lays down theſe
two things as certain ; that no Motion
muſt be allow'd the Earth, and that
God has made nothing in the Planets,
no not ſo much as Herbs, which has
either Life or Senſe in it. Leaving
then the Syſtem of *Copernicus*, he
chuſes *Tycho* for his Guide. But when
he ſuppoſes all the fix'd Stars to be
Suns, and round each of them places
their Planets, here (againſt his will I
ſuppoſe) he has unawares made an infinite number of *Copernican* Syſtems.
All which, befides their own Motion,
he abſurdly makes to be carry'd, with
a monſtrous ſwiftneſs, in twenty four
hours, round the Earth. When moſt
of theſe Worlds are out of the reach
of

of any Man's fight, as he owns they Book 2. are, I cannot think for what he makes fo many Suns to fhine upon defolate Lands (like our Earth in every thing, he fays, only that they have neither Plants nor Animals) where there's no one to whom they fhould give light. And from hence he ftill falls into more and more Abfurdities. And becaufe he could find no other ufe of the Planets, even in our Syftem, he is forc'd to beg help of the Aftrologers; and would have all thofe vaft Bodies made upon no other account than to preferve and rule the inferior World by, and govern the Mind of Man by their various and regular Influences. Accordingly, to gratify Aftrology, he fays that *Venus* was the prettieft pleafant place, every thing fine and handfom, its Light gentle, its Waters fweet and purling, and it felf befet all about with fhining Cryftals. In *Jupiter* he found wholefom and fweet Gales, delicate Waters, and a Land fhining like Silver. For from thefe two Planets forfooth, Men have all that is happy and healthful poured down upon them; and all that renders
them

them handſom and lovely, wiſe and grave, is owing to their Influences. *Mercury* had I don't know what ye call't, Airineſs and Briskneſs about him; whence Men derive, when they are firſt born, all their Wit and Cunning. *Mars* was nothing but deviliſh, infernal, ſtinking, black Flames and Smoke: and *Saturn* was all melancholy, dreadful, naſty, and dark: for theſe are the Planets (I don't know why, but all your Fortune-tellers hate them) that bring all the Plagues and Miſchiefs that we feel upon us, and would exerciſe their ſpite ſtill more, except they were ſometimes mitigated and correǎed by the benign and kind Influences of the other Planets. All this fine ſtuff his Genius teaches him. Which he makes give a ſerious Anſwer to this idle Queſtion, Whether a Jew or Heathen could be duly and rightly baptiz'd in the Waters of *Venus?* Of him too he learns that the Heaven of the fix'd Stars is no ſolid ſtuff, but a thin fluid, wherein an innumerable company of Stars and Suns lie floating here and there, not chain'd down to any place, (thus far he's in the

the right) and making in the fpace of Book 2.
a day that prodigious Tour round the
Earth. He forgets here, if there were
fuch a Motion, with what an incredible fwiftnefs they would fly out from
their Centers. But I fuppofe the In-
telligences that he has plac'd in them
will take care of that, thofe Angels
that prefide over, and regulate their
Motions. And in that he follows a
company of Doctors that harbour'd
that idle fancy of *Ariftotle* upon no account or confideration. But *Copernicus* has fet them all at liberty, only by
bringing in the Motion of the Earth :
which, if upon no other account, every one that is not blind purpofely muft
own to be neceffary upon this. I dare
fay *Kircher,* if he had dar'd freely to
fpeak his mind, could have afforded
us otherguefs things than thefe. But
when he could not have that liberty,
I think he might as well have let the
whole matter alone. But enough,
let's have done with this famous Author: And now that we have ventur'd to place Spectators in the Planets,
let's take a Journey to each of them,
and

and fee what their Years, Days, and Aftronomy are.

The Syftem of the Planets in Mercury. To begin with the innermoft and neareft the Sun : We know that *Mercury* is three times nearer that vaft body of Light than we are. Whence it follows that they fee him three times bigger, and feel him nine times hotter than we do. Such a degree of Heat would be intolerable to us, and fet a-fire all our dry'd Herbs, our Hay and Straw that we ufe. And yet I warrant the Animals there, are made of fuch a temper, as to be but moderately warm, and the Plants fuch as to be able to endure the Heat. The Inhabitants of *Mercury*, it's likely, have the fame opinion of us that we have of *Saturn*, that we muft be intolerably cold, and have little or no Light, we are fo far from the Sun. There's reafon to doubt, whether the *Mercurians*, tho they live fo much nearer the Sun, the Fountain of Life and Vigour, are much more airy and ingenious than we. For if we may guefs at them by what we fee here, we fhall not be obliged to grant it the

the Inhabitants of *Africa* and *Brasil*, Book 2.
that have got for their share the hot-
test places in the Earth, being neither
so wise nor so industrious as those that
belong to colder and more temperate
Climates ; they have scarce any Arts
or Knowlege among them, and those
of them that live upon the very shore,
understand little or no Navigation.
Nor can I be willing to make all that
vast number that must inhabit those
two large Planets, *Jupiter* and *Saturn*,
and have such noble Attendance, mere
dull Blockheads, or without as much
Wit as our selves, tho they are so far
more distant from the Sun. The Astronomy of the *Mercurials*, and the appearance of the Planets to them, oppo-
site at certain times to the Sun, may
be easily conceived by the Scheme of
the *Copernican* System in the former
Part. At the times of these Oppositi-
ons *Venus* and the Earth must needs
appear very bright and large to them.
For if *Venus* shines so gloriously to us
when she is new and horned, she must
necessarily in opposition to the Sun,
when she is full, be at least six or seven

Book 2. ven times larger, and a great deal
nearer to the Inhabitants of *Mercury*,
and afford them Light so strong and
bright, that they have no reason to
complain of their want of a Moon.
What the length of their Days are, or
whether they have different seasons in
the Year, is not yet discover'd, because we have not yet bin able to observe whether his Axis have any inclination to his Orbit, or what time he
spends in his diurnal Revolution upon
himself. And yet seeing *Mars*, the
Earth, *Jupiter* and *Saturn*, have certainly such Successions, there's no reason to doubt but that he has his Days
and Nights as well as they. But his
Year is scarce the fourth part so long as
ours.

In Venus.   The Inhabitants of *Venus* have
much the same face of things as those
in *Mercury*, only they never see him in
opposition to the Sun, which is occa-
sioned by his never removing above
38 degrees, or thereabouts, from it.
The Sun appears to them by half
larger in his Diameter, and above
twice in his Circumference, than to
us :

us : and by confequence affords them Book 2.
but twice as much Light and Heat, fo ᵥᵥᵥ
that they are nearer our Temperature
than *Mercury.* Their Year is com-
pleated in feven and a half of our
Months. In the Night our Earth,
when 'tis on the other fide of the Sun
from *Venus,* muft needs feem much
larger and lighter to *Venus* than fhe
doth ever to us ; and then they may
eafily fee, if they have not very weak
eyes, our conftant Attendant the Moon.
I have often wonder'd that when I
have viewed *Venus* at her neareft to
the Earth, when fhe refembled an
Half-moon, juft beginning to have
fomething like Horns, through a Telefcope of 45 or 60 Foot long, fhe always appeard to me all over equally
lucid, that I can't fay I obferv'd fo
much as one fpot in her, tho in *Jupi-*
ter and *Mars,* which feem much lefs
to us, they are very plainly perceived.
For if *Venus* had any fuch thing as Sea
and Land, the former muft neceffarily
fhow much more obfcure than the
other, as any one may fatisfy himfelf,
that from a very high Mountain will
but

Book 2. but look down upon our Earth. I thought that perhaps the too brisk Light of *Venus* might be the occasion of this equal appearance ; but when I used an Eye-glass that was smok'd for the purpose, it was still the same thing. What then, must *Venus* have no Sea, or do the Waters there reflect the Light more than ours do, or their Land less ? or rather (which is most probable in my opinion) is not all that Light we see reflected from an Atmosphere surrounding *Venus*, which being thicker and more solid than that in *Mars* or *Jupiter*, hinders our seeing any thing of the Globe it self, and is at the same time capable of sending back the Rays that it receives from the Sun ? For it's certain that if we look'd on the Earth from the outside of the Atmosphere, we should not perceive such a difference as we do from a Mountain; but by reason of the interposed Atmosphere, we should observe very little disparity between Sea and Land. 'Tis the same thing that hinders us from seeing the spots in the Moon as plain in the day as in the night,

night, becaufe the Vapors that fur- Book 2.
round the Earth being then enlightned 〰
by the Rays of the Sun, are an impediment to our profpect.

But *Mars*, as I faid before, has fome *In* Mars.
Parts of him darker than other fome.
By the conftant Returns of which his
Nights and Days have bin found to be
of about the fame length with ours.
But the Inhabitants have no perceivable difference between Summer and
Winter, the Axis of that Planet having
very little or no inclination to his Orbit, as has bin difcover'd by the Motion of his Spots. Our Earth muft appear to them almoft as *Venus* doth to
us, and by the help of a Telefcope
will be found to have its Wane, Increafe, and Full, like the Moon: and
never to remove from the Sun above
48 Degrees, by whofe difcovery they
fee it, as well as *Mercury* and *Venus*,
fometimes pafs. They as feldom fee
Venus as we do *Mercury*. I am apt to
believe, that the Land in *Mars* is of a
blacker hue than that of *Jupiter* or the
Moon, which is the reafon of his ap-
pearing of a Copper Colour, and his
reflect-

Book 2. reflecting a weaker Light than is proportionable to his diftance from the Sun. His Body, as I obferv'd before, tho farther from the Sun, is lefs than *Venus*. Nor has he any Moon to wait upon him, and in that, as well as *Mercury* and *Venus*, he muft acknowlege himfelf our inferiour. His Light and Heat is twice, and fometimes three times lefs than ours, to which I fuppofe the Conftitution of his Inhabitants is anfwerable.

Jupiter and Saturn the moft eminent of the Planets both for bignefs and attendants. If our Earth can claim preeminence of the fore-mentioned Planets, for having a Moon to attend upon it, (for its Magnitude can make but a fmall difference) how much fuperiour muft *Jupiter* and *Saturn* be to all four of them, Earth and all? For whether we confider their bulk, in which they far exceed all the others, or the number of Moons that wait upon them, it's very probable that they are the chief, the primary Planets in our Syftem, in comparifon with which the other four are nothing, and fcarce worth mentioning. For the eafier conception of their vaft difparity, I have thought fit

to

Fig. 3.

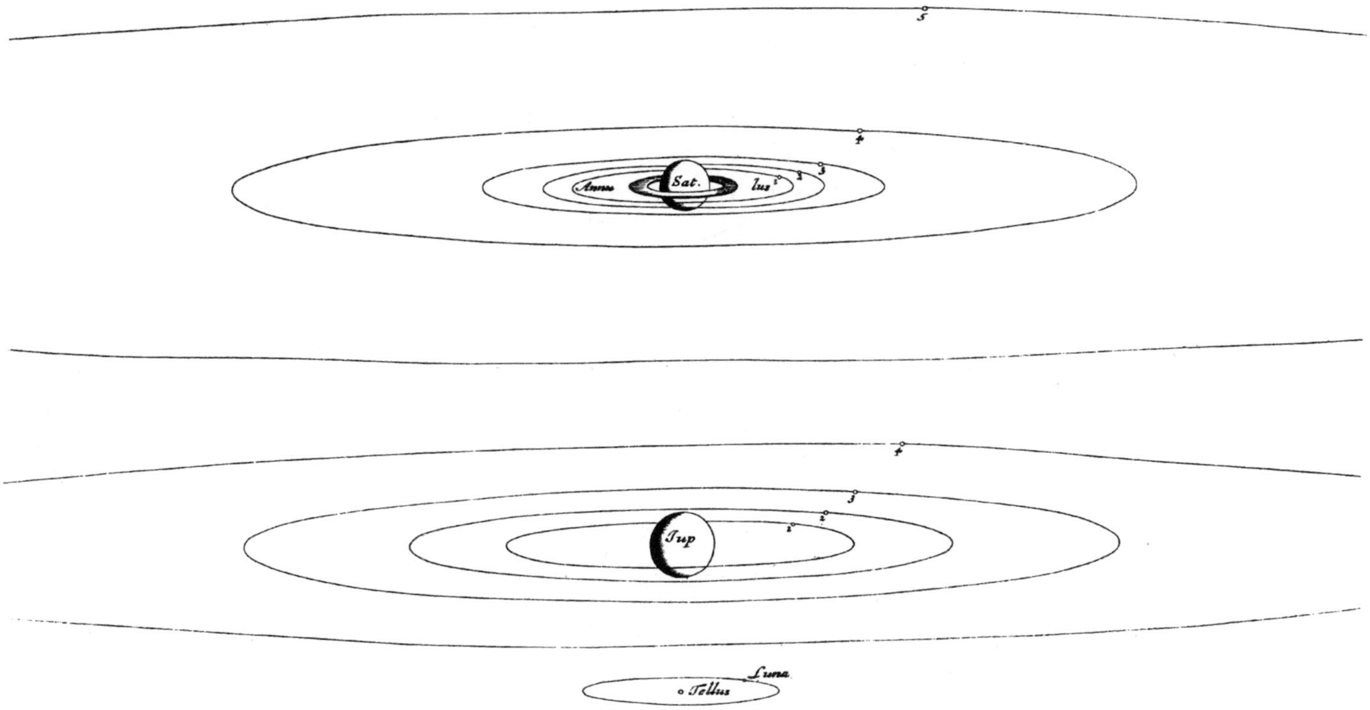

to add a Scheme of our Earth, with
the Path of the Moon about it, and
the Globe of the Moon it felf; and
the Syftems of *Jupiter* and *Saturn*,
where I have drawn every thing as *Fig.* 3.
near the true Proportion as poffible.
Jupiter you fee has his four, and *Sa-*
turn his five Moons about him, all
plac'd in their Orbits. The *Jovial* we
owe to *Galilæo*, 'tis well known : and
any one may imagine he was in no
fmall rapture at the difcovery. The
outermoft but one, and brighteft of
Saturn's, it chanc'd to be my lot, with
a Telefcope not above 12 foot long,
to have the firft fight of in the year
1655. The reft we may thank the in-
duftrious *Caffini* for, who ufed the
Glaffes of *Jof. Campanus*'s Work, firft
of 36, and afterwards of as many a-
bove 100 foot long. He has often,
and particularly in the year 1672,
fhow'd me the third and fifth. The
firft and fecond he gave me notice of
by Letters in the year 1684. but they
are fcarce ever to be feen, and I can't
pofitively fay I had ever that happinefs : but am as fatisfied that they are
there,

Book 2. there, as if I had ; not in the leaft fuf-
pecting the Credit of that worthy
man. Nay, I am afraid there are one
or two more ftill behind, and not without reafon. For between the fourth
and fifth there's a diftance not at all
proportionable to that between all the
others : Here for ought I know may
lurk a fixth Gentleman ; or perhaps
there may be another without the fifth
that may yet have efcaped us : for we
can never fee the fifth but in that part
of his Orbit, which is towards the
Weft : for which we fhall give you a
very good reafon.

Perhaps when *Saturn* comes into
the Northern Signs, and is at a good
height from the Horizon (for at the
writing of this he is at his loweft)
you may happen to make fome new
Difcoveries, good Brother, if you
would but make ufe of your two Telefcopes of 170 and 210 foot long ;
the longeft, and the beft I believe now
in the World. For tho we have not
yet had an opportunity of obferving
the Heavens with them (as well by
reafon of their Unweildinefs, as for
the

the interruption of our Studies by your abſence) yet I am ſatisfied of their Goodneſs by our trial of them one night, in reading a Letter at a vaſt diſtance by the help of a Light. I cannot but think of thoſe times with pleaſure, and of our diverting labour in poliſhing and preparing ſuch Glaſſes, in inventing new Methods and Engines, and always puſhing forward to ſtill greater and greater things. But to return to thoſe Diagrams.

I have there made the Diameter of *Jupiter* about two third parts of our diſtance from the Moon : for the Diameter of *Jupiter* is above twenty times bigger than that of the Earth ; which the diſtance of the Moon contains about thirty times. The Orbit of the outermoſt of *Jupiter*'s Guards is to that of the Moon round the Earth, as 8 and $\frac{1}{2}$ is to 1. And each of theſe Moons, by the ſhadow they make upon *Jupiter*, cannot be leſs than our Earth. Their Periods, that I may not omit them, are according to *Caſſini*'s account theſe. That of the inmoſt is one day, 18 hours, 28 minutes,

Book 2.

The proportion of the Diameter of Jupiter, and of the Orbs of his Satellites, to the Orbit of the Moon round the Earth.

The periods of Jupiter's Moons.

and

and 36 seconds. The second spends 3 days, 13 hours, 13 min. 52 sec. in going round him. The third 7 days, 3 hours, 59 min. 40 sec. The fourth 16 days, 18 hours, 5 min. 6 sec. The distance of the innermost from *Jupiter* himself is 2 ⅓ of his Diameters. That of the second is 4 and a half: Of the third 7 and one sixth part: Of the fourth 12 and two thirds, of the same Diameters. The innermost of *Sa-*

And Sa-turn's. *turn's* Guards moves round him in 1 day, 21 hours, 18 min. 31 sec. The second in 2 days, 17 hours, 41 min. 27 sec. The third in 4 days, 13 hours, 47 min. 16 sec. The fourth in 15 days, 22 hours, 41 min. 11 sec. The fifth in 79 days, 7 hours, 53 min. 57 sec. Their distances from the Center of *Saturn* are, that of the first almost one, that is 39 fortieth parts of the Diameter of his Ring; that of the second one and a quarter of those Diameters; of the third one and three quarters of them ; of the fourth four, or according to my calculation, but 3 and a half; of the fifth 12, which were found with vast pains and labour.

Now

Now can any one look upon, and
compare thefe Syftems together, without being amazed at the vaft Magnitude and noble Attendance of thefe
two Planets, in refpect of this little
pitiful Earth of ours? Or can they
force themfelves to think, that the wife
Creator has difpofed of all his Animals and Plants here, has furnifh'd and
adorn'd this Spot only, and has left all
thofe Worlds bare and deftitute of Inhabitants, who might adore and worfhip him; or that all thofe prodigious Bodies were made only to twinkle
to, and be ftudied by fome few perhaps of us poor fellows?

I do not doubt but there will be *This pro-*
fome who will think we Romance *portion true*
very much about the Magnitude of *according*
thefe Planets. For will you pretend *to all mo-*
to make them who are taken up in ad- *dern Obfer-*
miring the largenefs of this Globe, its *vations.*
multitude of Nations, Cities, and Em-
pires; can you pretend I fay to make
them ever believe that there are Places
in comparifon of which the Earth is as
inconfiderable as my Figure would
make it? No, they know better things
they'l

Book 2. they'l cry. But they may vouchfafe to be inform'd, that thefe Proportions are thofe which the beft Aftronomers of this Age have agreed upon. For if the Earth be diftant from the Sun ten or eleven thoufand of its own Diameters, according to the accounts of Monfieur *Caſſini* in *France*, and Mr. *Flamfted* in *England*, wherein they made ufe of very exact Obfervations of the Parallaxes of *Mars*; or if, according to a very probable Conjecture of mine, it be diftant twelve thoufand, then the Magnitudes of the other Orbs will very near anfwer the Proportions here fettled.

The apparent magnitude of the Sun in Jupiter, and a way of finding what light they there enjoy.   But to return to *Jupiter*. The Sun appears to them five times lefs than to us, and confequently they have but the five and twentieth part of the Light and Heat that we receive from it. But that Light is not fo weak as we imagine, as is plain by the brightnefs of that Planet in the Night; and that when the Sun is fo far eclips'd to us, as that the 2 5*th* part of his Difk be not free from the Shadow, he is not fenfibly darken'd. But if you have a mind

mind exactly to know the quantity of Book 2.
Light that *Jupiter* enjoys, you may take ◡◠◡
a Tube of what length you please. Let
one end of it be clos'd with a Plate of
Brafs, or any fuch thing, in the middle of which there muft be a hole,
whofe breadth muft have the fame
proportion to the length of the Tube,
as the Chord of 6 Minutes bears to the
Radius ; that is about as one is to 570.
Let the Tube be turn'd fo to the Sun,
that no Light may fall upon a white
Paper plac'd at the end of it, but what
comes through the little hole at the
other end of the Tube. The Rays
that come through this will reprefent
the Sun upon the Paper of the fame
Brightnefs that the Inhabitants of
Jupiter fee it in a clear day. And
if removing the Paper you place your
eye in the fame place, you will fee the
Sun of the fame Magnitude and
Brightnefs as you would were you in
Jupiter.

If you make the hole twice as little *And in Sa-*
in breadth, you will fee the fame of *turn.*
Saturn. And altho his Light be but
the hundredth part of ours, yet you
fee

Book 2. ſee it makes him ſhine finely in a dark
night. But in cloudy days what ſhall
the poor Inhabitants do? Why if we
were to be Judges but miſerably, but
yet I warrant they do not at all complain. Perhaps they may be like Owls
and Bats, and may love the Twilight
better than open day.

In Jupiter But it's a little ſtrange, that when
their days *Jupiter* is ſo much bigger than our Pla-
are 5 hours net, their Days and Nights ſhould be
but five of our Hours. By this we
may ſee that Nature has not obſerv'd
that proportion that their bulk ſeems
to require, ſeeing in *Mars* the days are
very little different from ours. But in
the length of their years, that is in the
revolution of the Planets round the
Sun, there is an exact proportion to
their diſtances from the Sun followed.
For as the Cubes of their diſtances, ſo
are the Squares of their Revolutions,
as *Kepler* firſt found out. Which proportion the Moons of *Jupiter* and *Sa-*
turn keep in their Courſes round thoſe
Always of Planets. As the Years and Days in
the ſame *Jupiter* are different from ours in this
length. reſpect, ſo are the Days in another ;
namely,

namely, that they are all of the fame Book 2.
length. For they there enjoy a perpetual Equinox, their Axis having little
or no inclination to their Orbit, as the
Earth's has, as has bin difcover'd by
Telefcopes. The Countries that lie
near their Poles have little or no heat,
by reafon the Rays of the Sun fall fo
obliquely upon them ; but then they
are freed from the Inconveniency that
ours are troubled with, of tedious long
half-year Nights, and have the conftant returns of Day and Night every
five hours. Indeed we fhould not be
contented with fuch fhort days, and
fhould count our felves very ill dealt
with if we had not twice as long, tho
upon no other account, but that what
is our own, to be fure, muft be beft.

The reft of the Planets are fo near
the Sun, (*Mars* himfelf never being
above 18 degrees from it) that in *Ju-*
piter they have the fight only of *Sa-*
turn. But we cannot deny but that
their four Moons ftand them in greater
ftead than our one doth us, if 'twere
only that they feldom know any fuch
thing as to be without Moonfhiny
Nights.

Book 2. Nights. And they are of great advantage to them, as we said before, in their Navigation, if they have any such thing. Not to mention the pleasant sights of their frequent Conjunctions and Eclipses, things that they are seldom a day without.

Saturn enjoys all those Pleasures and Advantages in a still higher degree, as well for his five Moons, as for the delightful prospect that the Ring about him affords his Inhabitants night and day. But we will be as kind to them as we have bin to the rest of the Planets, in giving an account of their Astronomy.

They see the fixt Stars just as we do. And first of all we shall observe what we might have remark'd before, but will be more strange here, that the fix'd Stars appear to them of the same Figure and Magnitude, and with the same degree of Light that they do to us : and this, by reason of their immense distance, of which we shall have occasion to speak by and by. In comparison with which the space that a Bullet shot out of a Cannon could travel in 25 years, would be almost nothing. Their

Their Aſtronomers have all the
ſame Signs of the Bear, the Lion, O-
rion, and the reſt, but not turning upon the ſame Axis with us: for that's
different in all the Planets.

As *Jupiter* can ſee no Planet but *Sa-*
turn, ſo *Saturn* knows of no Planet
but *Jupiter* ; which appears to him
much as *Venus* doth to us, never removing above 37 degrees from the
Sun. The length of their days I cannot determine : But if from the diſtance and period of his innermoſt Attendant, and comparing it with the
innermoſt of *Jupiter*'s, a Man may
venture to give a gueſs, they are very
little different from *Jupiter*'s, 10 hours
or ſomewhat leſs. But whereas in
Jupiter theſe are equally divided between Light and Darkneſs, the *Satur-*
nians muſt perceive a more ſenſible
difference than we, eſpecially between
Summer and Winter. For our Axis
inclines to the place of the Ecliptick
but 23 degrees and a half, but there's
above 31. Upon this account his
Moons muſt decline very much from
the Path that the Sun ſeems to move
in,

in, and his Inhabitants can never have a full Moon but juſt at the Equinoxes: two of which fall out in 30 of our years. 'Tis this Poſition of the Axis too that is the cauſe of thoſe delightful appearances, and wonderful proſpeᶜts that its Inhabitants enjoy: for the better underſtanding of which I ſhall draw a Figure of *Saturn* with his Ring about him: in which the proportion between the Diameters of the Globe and Ring is as 9 to 4. And the empty ſpace between them is of the ſame breadth with the Ring it ſelf. All Obſervations conſpire to prove that that is of no great thickneſs, altho if we ſhould allow it ſix hundred German Miles, I think, conſidering its Diameter, we ſhould not overdo the matter.

Fig. 4. Suppoſe then that to be the Globe of *Saturn*, whoſe Poles are A, B. G N is the Diameter of the Ring, as you view it ſideways, repreſenting a narrow Oval. Thoſe that live about the Poles within the Arches C A D, E B F, each of which are 54 degrees, (if the Cold will ſuffer any body to live there)

Fig. 4.

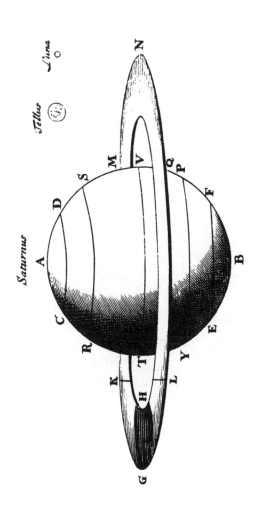

Saturnus

Tellus

Luna

there) never have a fight of the Ring. Book 2.
From all other parts it is continually to ∿
be seen for fourteen years and nine *The appea-*
Months, which is juft half their year. *rances of*
the Ring in
The other half it is hid from their Saturn.
view. Thofe then that dwell between
the Polar Circle C D, and the Equa-
tor T V, all that time that the Sun en-
lightens the part oppofite to them,
have every night the fight of a piece of
it H G L, much in the fhape of a
fhining Bow, which comes from the
Horizon, but is darken'd in the middle by the fhadow of *Saturn* G H,
which reaches moft commonly to the
outermoft rim of it. But after mid-
night that Shadow by little and little
begins to move towards the right hand
to thofe in the Northern, but the left
to thofe in the Southern Hemifphere.
In the morning it difappears, leaving
behind it a likenefs indeed of a Bow,
but much paler and weaker than our
Moon is in the day-time. For they,
as I faid before, have an Atmofphere,
or an Air furrounding them enlighten'd
by the Sun. Otherwife Night and
Day they would have their Ring,
their

Book 2. their Moons, and all the fix'd Stars, equally confpicuous. Another thing that muft make the fight of their Ring very curious, is, that by fome Spots in it, it is difcover'd to turn round upon it felf: A thing that thofe that are fo near cannot but take notice of, when we that live at this diftance can defcry a great Inequality, the infide of it being brighter much than the outfide is. When the fhadow of the Globe falls upon that part of the Ring G H, the fhadow of the Ring at the fame time darkens another part of the Globe about P F, which otherwife would have the Sun upon it. So that there is always a Zone of the Globe P Y F E, fometimes of a larger extent than at others, which is depriv'd of the fight both of the Sun and Ring for a confiderable time, the latter of which hides fome part of the Stars from it too. An amazing thing it muft be, all of a fudden to have the Sun darken'd, and fall into a pitch-night, without feeing any caufe of fuch an accident. All which while their Moons are their only Comfort. The other half of the

year

year the Hemisphere T B V enjoys the Book 2.
same Light that T A U before did,
and then this undergoes those long
Eclipses that that before suffer'd. At
the Equinoxes, when the Sun is in the
same Plane with the Ring, the Satur-
nians cannot well perceive it : no not
even we with our Glasses, by reason
of its Darkness. This happens when
Saturn, viewed from the Sun, is
advanced one and twenty degrees
and a half in *Virgo* or *Pisces*, as I
show'd formerly in my System of *Saturn*: Where there is an account given
of the Risings of the Sun above the
Ring, throughout all the Saturnian
Year.

With *Saturn* in this Scheme you
have the Globes of the Earth and
Moon drawn in their true proportion,
to put you in mind again of a thing ve
ry fit to be remember'd, how very
small our Habitation is when compar'd
with that Globe or the Ring about it.
And now any one, I suppose, can
frame to himself a picture of the
Night in *Saturn*, with two Arches of
the Ring, and five Moons shining
about,

Book 2. about, and adorning him. This then ſhall be what I have to ſay to the primary Planets.

We are now come a little lower, to make an enquiry into the Attendants of theſe Planets, eſpecially our own. And here we ſhall meddle not only with their Aſtronomy, but ſhall alſo ſearch into their Furniture and Ornament, if they are found to have any ſuch thing, which we have put off conſidering till now.

Very little to be ſaid of the Moon.

And here one would think that when the Moon is ſo near us, and by the means of a Teleſcope may be ſo nicely and exactly obſerv'd; it ſhould afford us matter for more probable Conjectures than any of the other remote Planets. But it is quite otherwiſe, and I can ſcarce find any thing to ſay of it, becauſe I have not a Planet of the ſame nature before my eyes, as in all the primary ones I have. For they are of the ſame kind with our Earth; and ſeeing all the Actions, and every thing that is here, we may make a reaſonable Conjecture at what we cannot ſee in thoſe Worlds.

But

But this we may venture to fay, Book 2.
without fear, that all the Attendants *The*
of *Jupiter* and *Saturn* are of the fame *Guards of*
nature with our Moon, as going round *Jupiter*
them, and being carry'd with them *and Sa-*
round the Sun juft as the Moon is with *turn of the*
the Earth. Their Likenefs reaches to *fame na-*
other things too, as you'l fee by and *ture with*
by. Therefore whatfoever we can *our Moon.*
with reafon affirm or fancy of our
Moon (and we may fay a little of it)
muft be fuppos'd with very little alteration to belong to the Guards of *Ju-*
piter and *Saturn*, as having no reafon
to be at all inferior to that.

The Surface of the Moon then is *The Moon*
found, by the leaft Telefcopes of about *hath Moun-*
three or four foot, to be diverfified *tains.*
with long Tracts of Mountains, and
again with broad Valleys. For in
thofe parts oppofite to the Sun you
may fee the Shadows of the Mountains, and often difcover the little
round Valleys between them, with a
hillock or two perhaps rifing out of
them. *Kepler* from the exact roundnefs of them would prove that they
are fome vaft work of the rational

Inha-

Book 2. Inhabitants. But I can't be of his mind, both for their incredible largeneſs, and that they might eaſily be occaſion'd by natural Cauſes. Nor can I find any thing like Sea there, tho he and many others are of the contrary opinion I know. For thoſe vaſt Countries which appear darker than the other, commonly taken for and call'd by the names of Seas, are diſcover'd with a good long Teleſcope, to be full of little round Cavities; whoſe Shadow falling within themſelves, makes them appear of that colour: and thoſe large Champains there in the Moon you will find not to be always even and ſmooth, if you look carefully *But no Sea,* upon them: neither of which two things can agree to the Sea. Therefore thoſe Plains in her that ſeem brighter than the other parts, muſt conſiſt, I ſuppoſe, of a whiter ſort of Matter than they. Nor do I believe *Nor Ri-* that there are any Rivers, for if there *vers,* were, they could never eſcape our ſight, eſpecially if they run between the Hills as ours do. Nor have they *NorClouds,* any Clouds to furniſh the Rivers with Water.

Water. For if they had, we fhould Book 2.
fometimes fee one part of the Moon
darken'd by them, and fometimes
another, whereas we have always the
fame profpect of her.

'Tis certain moreover, that the *Nor Air,*
and Water.
Moon has no Air or Atmofphere furrounding it as we have. For then we
could never fee the very outermoft
Rim of the Moon fo exactly as we do,
when any Star goes under it, but its
Light would terminate in a gradual
faint fhade, and there would be a fort
of a down as it were about it ; not to
mention, that the Vapors of our Atmofphere confift of Water, and confequently that where there are no Seas
or Rivers, there can be no Atmofphere. This is that notable difference
between that Planet and us that hinders all probable Conjectures about it.
If we could but once be fure that they
had Water, we might come to an A-
greement, and plant a Colony perhaps
there ; we might allow it then moft of
our other Privileges, and, with *Xeno-*
phanes, furnifh it with Inhabitants,
Cities, and Mountains. But as 'tis, I
cannot

Book 2. cannot imagine how any Plants or A-
nimals, whose whole nourishment
comes from liquid Bodies, can thrive
in a dry, waterless, parch'd Soil.

The Con-   What then, shall this great Ball be
jecture of made for, nothing but to give us a little
its Plants puny light in the Night-time, or to
and Ani- raise our Tides in the Sea? Shall not
mals very we plant some People there that may
dubious. have the pleasure of seeing our Earth
turn upon it self, presenting them sometimes with a prospect of *Europe* and
Africa, and then of *Asia* and *America* ;
sometimes half, and sometimes full?
What! and must all those Moons
round *Jupiter* and *Saturn* be con-
demn'd to the same Uselesness? I do
not know what to think of it, because
I know of nothing like them to found
a Conjecture upon. And yet 'tis not
improbable that those great and noble
Bodies have somewhat or other growing and living upon them, tho very
different from what we see and enjoy
here. Perhaps their Plants and Animals may have another sort of Nourishment there. Perhaps the moisture
of the Earth there is but just sufficient
to

to caufe a Mift or Dew, which may Book 2.
be very futable to the growth of their
Herbs. Which I remember is *Plu-*
tarch's opinion, in his Dialogue upon
this Subject. For in our Earth a very
little Water drawn from the Sea into
Dew, and falling down again upon
the Herbs, would be fufficient for all
our needs, without any Rain or Showers. But thefe are mere guefles, or
rather doubts, but yet they are the
beft we can make of this, and all
thofe other Moons: for, as I faid be- Jupiter's
fore, they are all of the fame nature, *and Sa-*
which is proved likewife by this, that *turn's*
Moons turn
as our Moon can afford us the fight ne- *always the*
ver but of one fide of her, fo they *fame fide*
to them.
turn always the fame face to their primary Planets. You wonder, I fuppofe, how we came to know fo much ;
but 'twas no hard matter, after that
Obfervation which I juft now made,
that the outermoft of *Saturn's* Moons
can never be feen but when fhe is on
the Weft-fide of her Planet. The
reafon of which is plainly this, that
one fide of her is darker, and does not
reflect the Light fo much as the other
which

which when it is turned towards us, we cannot see by reason of its weak Light. This always happening when 'tis East of him, and never on the other side, is a manifest proof that she always keeps the same side toward *Saturn*. Now since the outermost of *Saturn's* and our Moon carry themselves thus to the Planets round which they move, who can well doubt it of all the rest round *Jupiter* and *Saturn?* And there's a very good reason for it, namely, that the matter of which those Moons consist, being heavier, and more solid on the side that is averse from us, than on that which we have the sight of, does consequently fly with a greater force from the Centre of its Motion: for otherwise, according to the Laws of Motion, it should turn the same side always, not to its Planet, but to the same fixt Stars.

This Position of the Moons, in respect of their Planets, must occasion a great many very pretty, wonderful sights to their Inhabitants, if they have any: which is very doubtful, but may for the present be suppos'd.

An

An enquiry into our Moon may ferve Book 2.
for all the reft. Its Globe is divided
into two parts, after that manner,
that thofe who live on one fide never
lofe the fight of us, and thofe on the
other never enjoy it. Only thofe who
live on the Confines of each of thefe
lofe us, and fee us again by turns. The *The Aftro-*
Earth to them muft feem much larger *nomy of the*
than the Moon doth to us, as being in *tants of*
Diameter above four times bigger. *the Moon.*
But the beft of it is, that night and
day they fee it always in the very fame
part of the Heaven, as if it never
moved : fome of them as if 'twas falling upon their heads : others fomewhat above the Horizon, and others
always in the Horizon, ftill turning
upon it felf, and prefenting them every twenty four hours with a view of
all its Countries, even of thofe that
lie near the Poles (I could wifh my
felf in the Moon only for the fight of
them) yet unknown and undifcover'd
by us. They have it in its monthly
Wane and Increafe, they fee it half,
and horned, and full, by turns, juft as
we do their Planet. But the Light
that

Book 2. that they borrow of us is five times larger than what they pay us again. So that in dark nights that part that hath the advantage of being towards us, receives a very glorious Light, tho let *Kepler* fay what he will, no Heat from us. Their Days are always of the fame length with their Nights; and the Sun rifing and fetting to them but once in one of our Months, makes the time both of their Light and Darkneſs to be equal to 15 of our days. If their Bodies are of the fame Metal with ours, thoſe that have the Sun pretty high in their Horizon, muſt be like to be burnt up in fuch long days. For the Sun is not farther from them than he is from us. This will be the caſe of thoſe that live upon the Borders of the two Hemiſpheres we talk'd of; but thoſe that live under the Poles of the Moon will be juſt about as hot as our Whale-Fiſhers about *Iſland* and *Nova Zemla* are, in the Summer-time: who are in fo little danger of being roaſted, that in the middle of their Summer, in their days of three Months length, they are ready to loſe their fin-

fingers ends. The Poles of the Moon Book 2.
I call thofe, round which the fixt Stars ᗐᘯ
feem to turn to its Inhabitants, which
are different from ours, and thofe of
the Ecliptick, altho they move round
thefe latter, at the diftance of five degrees, in a period of nineteen Years.
Their Year they count by the Motion
of the Stars, and their return to the
Sun, and 'tis the fame with ours.
They can eafily do it, becaufe they
have the Stars day and night, notwithftanding the Light of the Sun:
for they have no Atmofphere (which
is the only reafon that we don't every
day enjoy the fame fight) to hinder
their Obfervations. Nor have they
any Clouds to obftruct their view, fo
that they have an eafier work than we
to find out the Courfes, but a more
difficult to make a true Syftem of the
Planets. For they will be apt to lay
a wrong Foundation upon the Immo-
bility of the Earth, which will lead
them into more dangerous Errors than *This may*
ever it did us. All that I have faid *be applied*
to the
belongs as well to *Jupiter*'s and *Sa-* *Moons a-*
turn's as to our Moon, in refpect of *bout Jupi-*
ter and
the *Saturn.*

Book 2. the Planets they move round. The
length of their Day and Night is always equal to the time of their Revolution: for example, the fifth Moon
moves round *Saturn* in 80 days, and
the days and nights there are equal to
forty of ours. Both their Summer
and Winter (*Saturn* moving round the
Sun in thirty years) are fifteen years
long. Therefore it is impossible but
that their way of living must be very
different from ours, having such tedious Winters, and such long watching
and sleeping times.

Having thus explain'd the primary
and secondary Planets round the Sun,
we should next set about the third sort,
the Sun and fixt Stars ; but before we
do that, it will be worth while to set
before you at once, in a clearer and
more plain Method than hitherto, the
Magnificence and Fabrick of the Solar
System. Which we can't possibly
do in so small a space as one of our
Leaves will but admit of, because the
Bodies of the Planets are so prodigi-
ously small in comparison of their Orbs.
But what is wanting in Figure shall be
made

Fig: 5.

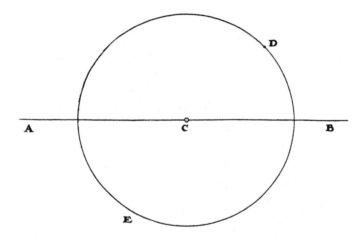

made up in Words. Going back then Book 2.
to the firſt Scheme, ſuppoſe another 〰
like it, and proportionable, drawn *Fig.* 1.
upon a very large ſmooth Plain;
whoſe outermoſt Circle repreſenting
the Orb of *Saturn*, muſt be conceived
three hundred and ſixty foot in Semidiameter. In which you muſt place
the Globe and Ring of *Saturn* of that
bigneſs as the 2*d* Figure ſhows you. *Fig.* 2.
Let all the other Planets be ſuppoſed
every one in his own Orbit, and in
the middle of all the Sun, of the ſame
bigneſs that that Figure repreſents,
namely, about four inches in Diameter. And then the Orbit or Circle in
which the Earth moves, which the
Aſtronomers call the *magnus Orbis*,
muſt have about ſix and thirty foot in
Semidiameter. In which the Earth
muſt be conceived moving, not bigger
than a grain of Millet, and her Com-
panion the Moon ſcarcely perceivable,
moving round her in a Circle a little
more than two Inches broad, as in the
Figure here adjoined, where the *Fig.* 5.
line A B repreſents a ſmall portion of
that Circle which the Earth moves in:
the

Book 2. the fmall Circle therein C is the Earth, and the Circle D E the path of the Moon round it, in which the body of the Moon is D.

The outermoft of *Saturn*'s Moons moves in an Orbit whofe Semidiameter is 29 inches; that of *Jupiter* in a fomewhat fmaller, whofe Semidiameter is 19 and a quarter.

And thus we have a true and exact Defcription of the Sun's Palace, where the Earth will be twelve thoufand of its Semidiameters diftant from him, which in German Miles makes above feventeen Millions. But perhaps we may have a clearer comprehenfion of this vaft length, by comparing it with fome very fwift Motion. 'Twas a pretty fancy of *Hefiod*, that an Anvil let fall from the top of Heaven, reach'd the Earth the tenth day of its Journey, and in ten more arriv'd at the bottom of Hell, the end of it: fo making the Earth the midway between Heaven and Hell. I fhan't make ufe of the Anvil, but of one as good, namely, a Bullet fhot out of a great Gun, which may travel perhaps

haps in a moment, or pulfe of an Ar-
tery, about a hundred Fathom, as is ∿
prov'd by thofe Experiments that
Merfennus in a Treatife of his relates;
wherein the Sound was found to extend
it felf eighty hundredth parts in that
time. I fay then, that fuppofing a Bullet *The im-*
to move with this fwiftnefs from the *menfe di-*
ftance be-
Earth to the Sun, it would fpend 25 *tween the*
years in its paffage. To make a Jour- *Sun and*
Planets il-
ney from *Jupiter* to the Sun, would *luftrated.*
require 125, and from *Saturn* thither
250 years. This account depends upon the meafure of the Earth's Diameter, which, according to the accu-
rate Obfervations of the French, is
6538594 times fix *Paris* feet, one degree being 57060 of that meafure.
This fhows us how vaft thofe Orbs
muft be, and how inconfiderable this
Earth, the Theatre upon which all our
mighty Defigns, all our Navigations,
and all our Wars are tranfacted, is
when compared to them. A very fit
Confideration, and matter of Reflection, for thofe Kings and Princes who
facrifice the Lives of fo many People,
only to flatter their Ambition in being
Mafters

Book 2. Masters of some pitiful corner of this small Spot. But to return to the matter in hand, now we have given you an account of the Sun's proportion to those Orbs and Bodies, we'll see what more we can say of him.

No grouud for Conje- Bure in the Sun. And there are some that have bin so civil, as to allow the Sun himself his Inhabitants. But upon what reason I cannot imagine, there being less ground for a probability in him than in the Moon. For we are not yet sure, whether he be a compact or liquid Globe; altho, if my account of Light be true, upon that account I should rather think him liquid: which his roundness and equal distribution of his Light to all parts are an Argument for. For that inequality on his Surface, which is discover'd by the Telescopes, (and that not always neither) which makes men fancy boiling Seas and belching Mountains of Fire, is nothing but the trembling Motion of the Vapors our Atmosphere is full of near

The Facu- læ in the Sun not ea- sily seen. the Earth; which is likewise the cause of the Stars twinkling. Nor could I ever have the luck to discern those bright

bright Spots they brag fo much of in the Sun as well as of his dark ones, tho the latter I have very often feen ; fo that with very good reafon I can doubt whether there's any fuch thing. For, in all the exact Obfervations, I could never find any fuch pretended to be feen any where but juft about his dark Spots ; and it is no great wonder that thofe Parts which are fo near the darker, fhould appear fomewhat brighter than the reft. That the Sun is extremely hot and firy, is beyond all difpute, and fuch Bodies as ours could not live one moment in fuch a Furnace. We muft make a new fort of Animals then, fuch as we have no Idea or Likenefs of among us, fuch as we can neither imagine nor conceive : which is as much as to fay, that truly we have nothing at all to fay. No doubt that glorious and vaft Body was made for fome noble end and ufe, and fram'd with excellent defign. And I think we all very well know and feel its Ufefulnefs in that effufion of Light and Heat to all the Planets round it ; in the prefervation and happinefs of all

Book 2.

By reafon of its Heat no Inhabitants like ours can live in the Sun.

living

Book 2. living Creatures, and that not only in
our Ball, but in thofe vaft Globes of
Jupiter and *Saturn*, not much inferior
to its own. Thefe are fuch great, fuch
wife ends, that it is not ftrange that
the Sun fhould have bin made, if it
had bin only upon their account. For,
as for *Kepler's* fancy, that he hath another Office, namely, to help on the
Motion of the Planets in their own
Orbs, by turning them round their
Axis, (which he would fain eftablifh
in his Epitome) I fhall give good Rea-
fons why I cannot affent to it.

The fix'd
Stars fo
many Suns.
Before the invention of Telefcopes,
it feem'd to contradict *Copernicus's*
Opinion, to make the Sun one of the
fix'd Stars. For the Stars of the firft
Magnitude being efteemed to be about
three minutes Diameter; and *Coper-*
nicus (obferving that tho the Earth
changed its place, they always kept
the fame diftance from us) having ventur'd to fay that the *Magnus Orbis* was
but a point in refpect of the Sphere in
which they were placed, it was a plain
confequence that every one of them
that appeared any thing bright, muft
be

be larger than the Path or Orbit of the Book 2.
Earth : which is very abſurd. This ∽∼
is the topping Argument that *Tycho*
Brahe ſet up againſt *Copernicus*. But
when the Teleſcopes ſhav'd them of
their fictitious Rays, and ſhow'd 'em to
us bare and naked (which they do beſt
when the Eye-glaſs is black'd with
Smoke) juſt like little ſhining Points,
then that difficulty vaniſhed, and the
Stars might ſtill be Suns. Which is
the more probable, becauſe their Light
is certainly their own : for it's impoſſible that ever the Sun ſhould ſend, or
they reflect it at ſuch a vaſt diſtance.
This is the opinion that commonly goes
along with *Copernicus*'s Syſtem. And
the Patrons of it do alſo with reaſon *They are*
ſuppoſe, that all theſe Stars are not in *the ſame*
the ſame Sphere, as well becauſe there's *Sphere.*
no Argument for it, as that the Sun,
which is one of them, cannot be brought
to this Rule. But it's more likely they
are ſcatter'd and diſperſed all over the
immenſe ſpaces of the Heaven, and are
as far diſtant perhaps from one another, as the neareſt of them are from
the Sun.

Here

Book 2.   Here again too I know *Kepler* is of
another opinion in his Epitome of *Copernicus*'s Syftem, that we mention'd
above.   For tho he agrees with us,
that the Stars are diffus'd through all
the vaft Profundity, yet he cannot allow that they have as large an empty
fpace about them as our Sun has.   For
at that rate, 'twas his opinion, we
fhould fee but very few, and thofe of
very different Magnitudes : *For, feeing the largeft of all appear fo fmall to us,
that we can fcarce obferve or meafure them
with our beft Inftruments ; how muft thofe
appear that are three or four times farther from us ? Why, fuppofing them no
larger than thefe, they muft feem three or
four times lefs, and fo on till a little farther they will not be to be feen at all :
Thus we fhall have the fight of but very
few Stars, and thofe very different one
from another ;* Whereas we have thou-
fands, and thofe not confiderably bigger or lefs than one another.   But this
by no means proves what he would
have it ; and his miftake was chiefly,
that he did not confider the nature of
Fire, which makes it be feen at fuch
distances,

diftances, and at fuch fmall Angles as all other Bodies would totally difap-pear under. A thing that we need go no farther than the Lamps fet along the Streets to prove. For altho they are a hundred foot from one another, yet you may count twenty of them in a continued row with your eyes, and yet the twentieth of them fcarce makes an Angle of fix Seconds. Cer-tainly then the glorious Light of the Stars muft do much more than this; fo that it's no wonder we fhould fee a thoufand or two of them with our bare eyes, and with a Telefcope dif-cover twenty times that number. But *Kepler* had a private defign in making the Sun thus fuperior to all the other Stars, and planting it in the middle of the World, attended with the Planets: a favor that he did not defire to grant the reft. For his aim was by it to ftrengthen his Cofmographical Myfte-ry, that the diftances of the Planets from the Sun are in a certain proporti-on to the Diameters of the Spheres that are infcrib'd within, and circumfcrib'd about *Euclid*'s Polyedrical Bodies.

Which

Which could never be fo much as probable, except there were but one Chorus of Planets moving round the Sun, and fo the Sun were the only one of his kind.

But that whole Myftery is nothing but an idle Dream taken from *Pythagoras* or *Plato*'s Philofophy. And the Author himfelf acknowleges that the Proportions do not agree fo well as they fhould, and is fain to invent two or three very filly excufes for it. And he ufes yet poorer Arguments to prove that the Univerfe is of a fpherical Figure, and that the number of the Stars muft neceffarily be finite, becaufe the Magnitude of each of them is fo. But what is worft of all is, that he fettles the fpace between the Sun and the concavity of the Sphere of the fix'd Stars, to be fix hundred thoufand of the Earth's Diameters. For this very good reafon, forfooth, that as the Diameter of the Sun is to that of the Orbit of *Saturn*, which he makes to be as 1 to 2000, fo is this Diameter to that of the Sphere of the fix'd Stars. A mere fancy without any fhadow of Reafon.

Reafon. I cannot but wonder how
fuch things as thefe could fall from fo
ingenious a Man, and fo great an A-
ftronomer. But I muft give my Vote,
with all the greateft Philofophers of
our Age, to have the Sun of the fame
nature with the fix'd Stars. And this
will give us a greater Idea of the
World, than all thofe other Opinions.
For then why may not every one of *The Stars*
thefe Stars or Suns have as great a Re- *have Pla-*
nets about
tinue as our Sun, of Planets, with *them like*
their Moons, to wait upon them? *our Sun.*
Nay there's a manifeft reafon why
they fhould. For let us fancy our
felves placed at an equal diftance from
the Sun and fix'd Stars; we fhould
then perceive no difference between
them. For, as for all the Planets that
we now fee attend the Sun, we fhould
not have the leaft glimpfe of them,
either that their Light would be too
weak to affect us, or that all the Orbs
in which they move would make up
one lucid point with the Sun. In this
ftation we fhould have no occafion to
imagine any difference between the
Stars, and fhould make no doubt if we
had

Book 2. had but the fight, and knew the nature of one of them, to make that the
Standard of all the reft. We are then
plac'd near one of them, namely, our
Sun, and fo near as to difcover fix other
Globes moving round him, fome of
them having others performing them
the fame Office. Why then fhall not
we make ufe of the fame Judgment
that we would in that cafe; and conclude, that our Star has no better attendance than the others? So that
what we allow'd the Planets, upon
the account of our enjoying it, we
muft likewife grant to all thofe Planets
that furround that prodigious number
of Suns. They muft have their Plants
and Animals, nay and their rational
ones too, and thofe as great Admirers,
and as diligent Obfervers of the Heavens as our felves; and muft confequently enjoy whatfoever is fubfervient to, and requifit for fuch Knowlege.

What a wonderful and amazing
Scheme have we here of the magnificent Vaftnefs of the Univerfe! So many Suns, fo many Earths, and every
one of them ftock'd with fo many
Herbs,

Herbs, Trees and Animals, and a-Book 2.
dorn'd with fo many Seas and Mountains! And how muft our wonder and
admiration be encreafed when we confider the prodigious diftance and multitude of the Stars?

That their diftance is fo immenfe,
that the fpace between the Earth and
Sun (which is no lefs than twelve
thoufand of the former's Diameters)
is almoft nothing when compar'd to
it, has more Proofs than one to confirm it. And this among the reft. If
you obferve two Stars near one another, as for example thofe in the middle of the Great Bears Tail, differing
very much from one another in Clearnefs, notwithftanding our changing
our Pofition in our Annual Orbit round
the Sun, and that there would be a
Parallax were the Star which is
brighter nearer us than the other, as is
very probable it is, yet whatever
part of the year you look upon them,
they will not in the leaft have altered
their diftance. Thofe that have hitherto undertook to calculate their Diftance, have not bin able perfe&ly to
com-

Book 2. compafs their defign, by reafon of the
extreme nicenefs and almoft impoffibi-
lity of the Obfervations requifite for
their purpofe. The only Method that
I fee remaining, to come at any tolerable probability in fo difficult a cafe, I
ſt all here make ufe of. Seeing then
that the Stars, as I faid before, are fo
many Suns, if we do but fuppofe one
of them equal to ours, it will follow
that its diftance from us is as much
greater than that of the Sun, as its apparent Diameter is lefs than the Diameter of the Sun. But the Stars, even
thofe of the firft Magnitude, tho
view'd through a Telefcope, are fo
very fmall that they feem only like fo
many fhining Points, without any perceivable breadth. So that fuch Obfervations can here do us no good. When
A way of I faw this would not fucceed, I ftudied
making a by what way I could fo leffen the Dia-
probable
guefs at the meter of the Sun, as to make it not
diftance of appear larger than the Dog, or any,
the Stars. other of the chief Stars. To this purpofe I clos'd one end of my twelve-
foot Tube with a very thin Plate, in
the middle of which I made a hole not
ex-

exceeding the twelfth part of a Line, Book 2. that is the hundred and forty fourth part of an Inch. That end I turn'd to the Sun, placing my Eye at the other, and I could fee fo much of the Sun as was in Diameter about the 182d part of the whole. But ftill that little piece of him was brighter much than the Dog-Star is in the cleareft night. I faw that this would not do, but that I muft leffen the Diameter of the Sun a great deal more. I made then fuch another hole in a Plate, and againft it I plac'd a little round Glafs that I had made ufe of in my Microfcopes, of much about the fame Diameter with the former hole. Then looking again towards the Sun (taking care that no Light might come near my eye to hinder my Obfervation) I found it appear'd of much the fame Clearnefs with *Sirius*. But cafting up my account, according to the Rules of *Dioptricks*, I found his Diameter now was but $\frac{1}{152}$ part of that hundred and eighty fecond part of his whole Diameter that I faw through the former hole. Multiplying $\frac{1}{152}$ and $\frac{1}{182}$ into one

Book 2. one another, the Product I found to
be $\frac{1}{27664}$. The Sun therefore being
contracted into such a compass, or being removed so far from us (for it's the
same thing) as to make his Diameter
but the 27664 part of that we every
day see, will send us just the same
Light as the Dog-star now doth.
And his distance then from us will be
to his present distance undoubtedly as
27664 is to 1 ; and his Diameter little above four thirds, 4‴.   Seeing
then *Sirius* is supposed equal to the
Sun, it follows that his Diameter is
likewise 4‴, and that his distance to
the distance of the Sun from us is as
27664 to 1.   And what an incredible distance that is, will appear by the
same way of reasoning that we used in
measuring that of the Sun.   For if
25 years are required for a Bullet out
of a Cannon, with its utmost Swift-
ness, to travel from the Sun to us;
then by multiplying the number
27664 into 25, we shall find that such
a Bullet would spend almost seven
hundred thousand years in its Journey
between us and the nearest of the fix'd
Stars.

Stars. And yet when in a clear night Book 2. we look upon them, we cannot think them above some few miles over our heads. What I have here enquir'd into, is concerning the nearest of them. And what a prodigious number must there be besides of those which are placed so deep in the vast spaces of Heaven, as to be as remote from these as these are from the Sun ! For if with our bare Eye we can observe above a thousand, and with a Telescope can discover ten or twenty times as many ; what bounds of number must we set to those which are out of the reach even of these Assistances! especially if we consider the infinite Power of God. Really, when I have bin reflecting thus with my self, methoughts all our Arithmetick was nothing, and we are vers'd but in the very Rudiments of Numbers, in comparison of this great Sum. For this requires an immense Treasury, not of twenty or thirty Figures only, in our decuple Progression, but of as many as there are Grains of Sand upon the shore. And yet who can say, that even

Book 2. even this number exceeds that of the
Fixt Stars? Some of the Antients, and
Jordanus Brunus. carry'd it further, in
declaring the Number infinite: he
would perfwade us that he has prov'd
it by many Arguments, tho in my opinion they are none of them conclufive.
Not that I think the contrary can ever
be made out. Indeed it feems to me
certain, that the Univerfe is infinitely
extended; but what God has bin pleas'd
to place beyond the Region of the
Stars, is as much above our Knowlege, as it is our Habitation.

Or what if beyond fuch a determinate fpace he has left an infinite Va-
cuum; to fhow, how inconfiderable
all that he has made is, to what his
Power could, had he fo pleas'd, have
produc'd? But I am falling, before I
am aware, into that intricate Difpute
of Infinity: Therefore I fhall wave
this, and not, as foon as I am free of
one, take upon me another difficult
Task. All that I fhall do more is to
add fomewhat of my opinion concerning the World, as it is a place for the
reception of the Suns or fix'd Stars,
                                    every

every one of which, I have ſhow'd Book 2.
may have their Planetary Syſtems
about them.

I am of opinion then that every Sun *Every Sun*
is ſurrounded with a Whirl-pool or *has a Vor-*
tex round
Vortex of Matter in a very ſwift Mo- *it, very*
tion ; tho not in the leaſt like *Cartes*'s *different*
either in their bulk, or manner of Mo- *from thoſe*
of Cartes.
tion. For *Cartes* makes his ſo large,
as every one of them to touch all the
others round them, in a flat Surface,
juſt as you have ſeen the Bladders that
Boys blow up in Soap-ſuds do : and
would have the whole Vortex to move
round the ſame way. But the Angles of every Vortex will be no ſmall
hindrance to ſuch a Motion. Then
the whole matter moving round at
once, upon the Axis as it were of a
Cylinder, did not a little puzzle him
in giving Reaſons for the Roundneſs
of the Sun : which however they may
ſatisfy ſome People that do not conſider them, really prove nothing of the
matter. In this æthereal matter the
Planets float, and are carry'd round by
its motion : and the thing that keeps
them in their own Orbs is, that they
them-

Book 2. themfelves, and the matter in which they fwim, equally ftrive to fly out from the Center of this Motion. Againft all which there are many Aftronomical Objections, fome of which I touch'd upon in my Effay of the Caufes of Gravity. Where I gave another account of the Planets not deferting their own Orbs; which is their Gravitation towards the Sun. I fhow'd there the Caufes of that Gravitation, and cannot but wonder that *Cartes*, the firft man that ever began to talk reafonably of that matter, fhould never meddle with, or light on it. *Plutarch* in his Book of the Moon abovementioned fays, that fome of the Antients were of opinion, that the reafon of the Moon's keeping her Orbit was, that the force of her Circular Motion was exactly equal to her Gravity, the one of which pull'd her to, as much as the other forc'd her off from the Centre. And in our Age *Alphonfus Borellus*, who was of this fame opinion in the other Planets as well as the Moon, makes the Gravitation of the primary Planets to be towards the Sun, as that of the fecondary is towards the Planets round which they move: Which Mr. *Ifaac Newton* has more fully explained, with a great deal of pains and fubtilty; and how from that caufe proceeds the Ellipticity of the Orbs of the Planets, found out by *Kepler*. According to my Notion of the Gravitation

vitation of the Planets to the Sun, the mat-Book 2.
ter of his Vortex muſt not all move the
ſame way, but after ſuch a manner as to
have its parts carry'd different ways on all
ſides. And yet there is no fear of its being
deſtroy'd by ſuch an irregular motion, becauſe the Æther round it, which is at reſt,
keeps the parts of it from flying out. With
the help of ſuch a Vortex as this I have pretended in that Eſſay to explain the Gravity
of Bodies on this Earth, and all the effects
of it. And I ſuppoſe there may be the
ſame cauſe as well of the Gravitation of the
Planets, and of our Earth among the reſt,
towards the Sun, as of their Roundneſs : a
thing ſo very hard to give an account of in
Cartes's Syſtem.

I muſt differ from him too in the bigneſs
of the Vortices, for I cannot allow them to
be ſo large as he would make them. I
would have them diſpers'd all about the immenſe ſpace, like ſo many little Whirl-
pools of Water, that one makes by the
ſtirring of a ſtick in any large Pond or Ri-
ver, a great way diſtant from one another.
And as their motions do not all intermix or
communicate with one another ; ſo in my
opinion muſt the Vortices of Stars be plac'd
as not to hinder one anothers free Circum-
rotations.

So that we may be ſecure, and never fear
that they will ſwallow up or deſtroy one
another ; for that was a mere fancy of *Car-*
tes's,

Book 2. *tes*'s, when he was a showing how a fix'd Star or Sun might be turn'd into a Planet. And 'tis plain, that when he writ it, he had no thoughts of the immense distance of the Stars from one another; particularly, by this one thing, that he would have a Comet as soon as ever it comes into our Vortex, to be seen by us. Which is as absurd as can be. For how could a Star, which gives us such a vast Light only from the Reflection of the Beams of the Sun, as he himself owns they do; how I say could that be so plainly seen at a distance ten thousand times larger than the Diameter of the Earth's Orbit? He could not but know that all round the Sun there is a vast Extensum; so vast, that in *Copernicus*'s System the *magnus Orbis* is counted but a point in comparison with it. But indeed all the whole story of Comets and Planets, and the Production of the World, is founded upon such poor and trifling grounds, that I have often wonder'd how an ingenious man could spend all that pains in making such fancies hang together. For my part, I shall be very well contented, and shall count I have done a great matter, if I can but come to any knowlege of the nature of things, as they now are, never troubling my head about their beginning, or how they were made, knowing that to be out of the reach of human Knowlege, or even Conjecture.

F I N I S,

INDEX

INDEX